The Standard
PILOT LOG

ASA-SP-40

The Standard Pilot Log
SP-40

© 2017 Aviation Supplies & Academics, Inc.

ASA-SP-40
ISBN 978-1-56027-329-5

Published by
Aviation Supplies & Academics, Inc.
7005 132nd Place SE
Newcastle, WA 98059-3153
Website: www.asa2fly.com
Email: asa@asa2fly.com

Printed in the China

18

TRANSPORTATION   USD $12.95

ISBN 978-1-56027-329-5

51295 >

9 781560 273295

The Standard Pilot Log

Name _____ **Logbook Number** _____

Mailing Address _____ **From** _____

_____ **To** _____

Certificates Held			Rating Record	Aircraft Type Ratings
Type	**Date Issued**	**Certificate Number**	☐ AIRPLANE ☐ INSTRUMENT ☐ UNMANNED (UAS)	_____
REMOTE PILOT	_____	_____	☐ SINGLE ENGINE: ☐ LAND ☐ SEA	_____
STUDENT PILOT	_____	_____	☐ MULTI-ENGINE: ☐ LAND ☐ SEA	_____
SPORT PILOT	_____	_____	☐ MULTI-ENGINE: Limited to Center Line Thrust	_____
RECREATIONAL PILOT	_____	_____	☐ GLIDER ☐ AERO TOW ☐ WINCH TOW	_____
PRIVATE PILOT	_____	_____	☐ POWERED PARACHUTE ☐ WEIGHT-SHIFT CONTROL	_____
COMMERCIAL PILOT	_____	_____	☐ ROTORCRAFT ☐ HELICOPTER ☐ GYROPLANE	_____
FLIGHT INSTRUCTOR	_____	_____	☐ LIGHTER-THAN-AIR ☐ AIRSHIP ☐ BALLOON	_____
AIRLINE TRANSPORT PILOT	_____	_____	☐ POWERED LIFT	_____
AVIATION TECHNICIAN	_____	_____	☐ GI BASIC ☐ GI ADVANCED ☐ GI INSTRUMENT	_____
GROUND INSTRUCTOR	_____	_____	☐ OTHER RATINGS (Specify) _____	_____

| YEAR 20___ DATE | AIRCRAFT MAKE & MODEL | AIRCRAFT IDENT. | POINTS OF DEPARTURE & ARRIVAL | | AIRCRAFT CATEGORY | | | | GROUND TRAINER | TYPE OF PILOTING TIME | | | |
			FROM	TO	AIRPLANE SEL	AIRPLANE MEL				DUAL RECEIVED	PILOT-IN-COMMAND		
			PAGE TOTAL										
			AMOUNT FORWARD										
			TOTAL TO DATE										

CONDITIONS OF FLIGHT										NO. INSTR. APPR.	NO. LDG. DAY / NIGHT	TOTAL DURATION OF FLIGHT		REMARKS, PROCEDURES, MANEUVERS
DAY		NIGHT		CROSS-COUNTRY		ACTUAL INSTR.		SIMULATED INSTR.						
														I certify that the statements made by me on this form are true.
														PILOT'S SIGNATURE

YEAR 20___ DATE	AIRCRAFT MAKE & MODEL	AIRCRAFT IDENT.	POINTS OF DEPARTURE & ARRIVAL		AIRCRAFT CATEGORY						GROUND TRAINER		TYPE OF PILOTING TIME					
			FROM	TO	AIRPLANE SEL		AIRPLANE MEL						DUAL RECEIVED		PILOT-IN-COMMAND			
				PAGE TOTAL														
				AMOUNT FORWARD														
				TOTAL TO DATE														

CONDITIONS OF FLIGHT										NO. INSTR. APPR.	NO. LDG. DAY / NIGHT		TOTAL DURATION OF FLIGHT		REMARKS, PROCEDURES, MANEUVERS
DAY		NIGHT		CROSS-COUNTRY		ACTUAL INSTR.		SIMULATED INSTR.							
															I certify that the statements made by me on this form are true.
															PILOT'S SIGNATURE

| YEAR 20___ DATE | AIRCRAFT MAKE & MODEL | AIRCRAFT IDENT. | POINTS OF DEPARTURE & ARRIVAL | | AIRCRAFT CATEGORY | | | | GROUND TRAINER | TYPE OF PILOTING TIME | | | | | |
			FROM	TO	AIRPLANE SEL		AIRPLANE MEL			DUAL RECEIVED		PILOT-IN-COMMAND			
PAGE TOTAL															
AMOUNT FORWARD															
TOTAL TO DATE															

CONDITIONS OF FLIGHT						NO. INSTR. APPR.	NO. LDG. DAY / NIGHT		TOTAL DURATION OF FLIGHT		REMARKS, PROCEDURES, MANEUVERS
DAY		NIGHT		CROSS-COUNTRY	ACTUAL INSTR.	SIMULATED INSTR.					

														I certify that the statements made by me on this form are true.
														PILOT'S SIGNATURE

YEAR 20___ DATE	AIRCRAFT MAKE & MODEL	AIRCRAFT IDENT.	POINTS OF DEPARTURE & ARRIVAL		AIRCRAFT CATEGORY				GROUND TRAINER		TYPE OF PILOTING TIME			
			FROM	TO	AIRPLANE SEL	AIRPLANE MEL					DUAL RECEIVED	PILOT-IN-COMMAND		
PAGE TOTAL														
AMOUNT FORWARD														
TOTAL TO DATE														

CONDITIONS OF FLIGHT												NO. INSTR. APPR.	NO. LDG. DAY / NIGHT		TOTAL DURATION OF FLIGHT		REMARKS, PROCEDURES, MANEUVERS
DAY		NIGHT		CROSS-COUNTRY		ACTUAL INSTR.		SIMULATED INSTR.									
																	I certify that the statements made by me on this form are true.
																	PILOT'S SIGNATURE

YEAR 20___ DATE	AIRCRAFT MAKE & MODEL	AIRCRAFT IDENT.	POINTS OF DEPARTURE & ARRIVAL		AIRCRAFT CATEGORY			GROUND TRAINER	TYPE OF PILOTING TIME			
			FROM	TO	AIRPLANE SEL	AIRPLANE MEL			DUAL RECEIVED	PILOT-IN-COMMAND		
PAGE TOTAL												
AMOUNT FORWARD												
TOTAL TO DATE												

CONDITIONS OF FLIGHT										NO. INSTR. APPR.	NO. LDG. DAY / NIGHT	TOTAL DURATION OF FLIGHT		REMARKS, PROCEDURES, MANEUVERS
DAY		NIGHT		CROSS-COUNTRY		ACTUAL INSTR.		SIMULATED INSTR.						
														I certify that the statements made by me on this form are true.
														PILOT'S SIGNATURE

YEAR 20___ DATE	AIRCRAFT MAKE & MODEL	AIRCRAFT IDENT.	POINTS OF DEPARTURE & ARRIVAL		AIRCRAFT CATEGORY			GROUND TRAINER	TYPE OF PILOTING TIME		
			FROM	TO	AIRPLANE SEL	AIRPLANE MEL			DUAL RECEIVED	PILOT-IN-COMMAND	
			PAGE TOTAL								
			AMOUNT FORWARD								
			TOTAL TO DATE								

CONDITIONS OF FLIGHT										NO. INSTR. APPR.	NO. LDG. DAY / NIGHT	TOTAL DURATION OF FLIGHT		REMARKS, PROCEDURES, MANEUVERS
DAY		NIGHT		CROSS-COUNTRY		ACTUAL INSTR.		SIMULATED INSTR.						
														I certify that the statements made by me on this form are true.
														PILOT'S SIGNATURE

| YEAR 20___ DATE | AIRCRAFT MAKE & MODEL | AIRCRAFT IDENT. | POINTS OF DEPARTURE & ARRIVAL | | AIRCRAFT CATEGORY | | | | | GROUND TRAINER | | TYPE OF PILOTING TIME | | | | | |
			FROM	TO	AIRPLANE SEL		AIRPLANE MEL					DUAL RECEIVED		PILOT-IN-COMMAND			
			PAGE TOTAL														
			AMOUNT FORWARD														
			TOTAL TO DATE														

CONDITIONS OF FLIGHT												NO. INSTR. APPR.	NO. LDG. DAY/NIGHT	TOTAL DURATION OF FLIGHT		REMARKS, PROCEDURES, MANEUVERS
DAY		NIGHT		CROSS-COUNTRY		ACTUAL INSTR.		SIMULATED INSTR.								
																I certify that the statements made by me on this form are true.
																PILOT'S SIGNATURE

| YEAR 20___ DATE | AIRCRAFT MAKE & MODEL | AIRCRAFT IDENT. | POINTS OF DEPARTURE & ARRIVAL | | AIRCRAFT CATEGORY | | | | | GROUND TRAINER | TYPE OF PILOTING TIME | | | | | |
			FROM	TO	AIRPLANE SEL		AIRPLANE MEL					DUAL RECEIVED		PILOT-IN-COMMAND		
			PAGE TOTAL													
			AMOUNT FORWARD													
			TOTAL TO DATE													

CONDITIONS OF FLIGHT										NO. INSTR. APPR.	NO. LDG. DAY / NIGHT	TOTAL DURATION OF FLIGHT		REMARKS, PROCEDURES, MANEUVERS
DAY		NIGHT		CROSS-COUNTRY		ACTUAL INSTR.		SIMULATED INSTR.						
														I certify that the statements made by me on this form are true.
														PILOT'S SIGNATURE

YEAR 20___ DATE	AIRCRAFT MAKE & MODEL	AIRCRAFT IDENT.	POINTS OF DEPARTURE & ARRIVAL		AIRCRAFT CATEGORY					GROUND TRAINER		TYPE OF PILOTING TIME					
			FROM	TO	AIRPLANE SEL		AIRPLANE MEL					DUAL RECEIVED		PILOT-IN-COMMAND			
				PAGE TOTAL													
				AMOUNT FORWARD													
				TOTAL TO DATE													

CONDITIONS OF FLIGHT										NO. INSTR. APPR.	NO. LDG. DAY / NIGHT	TOTAL DURATION OF FLIGHT		REMARKS, PROCEDURES, MANEUVERS
DAY		NIGHT		CROSS-COUNTRY		ACTUAL INSTR.		SIMULATED INSTR.						
														I certify that the statements made by me on this form are true.
														PILOT'S SIGNATURE

YEAR 20___ DATE	AIRCRAFT MAKE & MODEL	AIRCRAFT IDENT.	POINTS OF DEPARTURE & ARRIVAL		AIRCRAFT CATEGORY				GROUND TRAINER	TYPE OF PILOTING TIME			
			FROM	TO	AIRPLANE SEL	AIRPLANE MEL				DUAL RECEIVED	PILOT-IN-COMMAND		
			PAGE TOTAL										
			AMOUNT FORWARD										
			TOTAL TO DATE										

CONDITIONS OF FLIGHT										NO. INSTR. APPR.	NO. LDG.		TOTAL DURATION OF FLIGHT		REMARKS, PROCEDURES, MANEUVERS
DAY		NIGHT		CROSS-COUNTRY		ACTUAL INSTR.		SIMULATED INSTR.			DAY / NIGHT				
											/				
											/				
											/				
											/				
											/				
											/				
											/				
											/				*I certify that the statements made by me on this form are true.*
											/				
											/				PILOT'S SIGNATURE

YEAR 20___ DATE	AIRCRAFT MAKE & MODEL	AIRCRAFT IDENT.	POINTS OF DEPARTURE & ARRIVAL		AIRCRAFT CATEGORY					GROUND TRAINER	TYPE OF PILOTING TIME					
			FROM	TO	AIRPLANE SEL		AIRPLANE MEL				DUAL RECEIVED		PILOT-IN-COMMAND			
			PAGE TOTAL													
			AMOUNT FORWARD													
			TOTAL TO DATE													

CONDITIONS OF FLIGHT										NO. INSTR. APPR.	NO. LDG. DAY/NIGHT	TOTAL DURATION OF FLIGHT		REMARKS, PROCEDURES, MANEUVERS
DAY		NIGHT		CROSS-COUNTRY		ACTUAL INSTR.		SIMULATED INSTR.						
														I certify that the statements made by me on this form are true.
														PILOT'S SIGNATURE

YEAR 20___ DATE	AIRCRAFT MAKE & MODEL	AIRCRAFT IDENT.	POINTS OF DEPARTURE & ARRIVAL		AIRCRAFT CATEGORY					GROUND TRAINER	TYPE OF PILOTING TIME					
			FROM	TO	AIRPLANE SEL		AIRPLANE MEL				DUAL RECEIVED		PILOT-IN-COMMAND			
PAGE TOTAL																
AMOUNT FORWARD																
TOTAL TO DATE																

CONDITIONS OF FLIGHT										NO. INSTR. APPR.	NO. LDG. DAY/NIGHT	TOTAL DURATION OF FLIGHT		REMARKS, PROCEDURES, MANEUVERS
DAY		NIGHT		CROSS-COUNTRY		ACTUAL INSTR.		SIMULATED INSTR.						
														I certify that the statements made by me on this form are true.
														PILOT'S SIGNATURE

| YEAR 20__ DATE | AIRCRAFT MAKE & MODEL | AIRCRAFT IDENT. | POINTS OF DEPARTURE & ARRIVAL | | AIRCRAFT CATEGORY | | | | | GROUND TRAINER | TYPE OF PILOTING TIME | | | | | |
			FROM	TO	AIRPLANE SEL		AIRPLANE MEL				DUAL RECEIVED		PILOT-IN-COMMAND			
			PAGE TOTAL													
			AMOUNT FORWARD													
			TOTAL TO DATE													

CONDITIONS OF FLIGHT											NO. INSTR. APPR.	NO. LDG. DAY / NIGHT	TOTAL DURATION OF FLIGHT		REMARKS, PROCEDURES, MANEUVERS
DAY		NIGHT		CROSS-COUNTRY		ACTUAL INSTR.		SIMULATED INSTR.							
															I certify that the statements made by me on this form are true.
															PILOT'S SIGNATURE

YEAR 20___ DATE	AIRCRAFT MAKE & MODEL	AIRCRAFT IDENT.	POINTS OF DEPARTURE & ARRIVAL		AIRCRAFT CATEGORY					GROUND TRAINER		TYPE OF PILOTING TIME					
			FROM	TO	AIRPLANE SEL		AIRPLANE MEL					DUAL RECEIVED		PILOT-IN-COMMAND			
		PAGE TOTAL															
		AMOUNT FORWARD															
		TOTAL TO DATE															

CONDITIONS OF FLIGHT										NO. INSTR. APPR.	NO. LDG. DAY/NIGHT		TOTAL DURATION OF FLIGHT		REMARKS, PROCEDURES, MANEUVERS
DAY		NIGHT		CROSS-COUNTRY		ACTUAL INSTR.		SIMULATED INSTR.							
															I certify that the statements made by me on this form are true.
															PILOT'S SIGNATURE

YEAR 20___ DATE	AIRCRAFT MAKE & MODEL	AIRCRAFT IDENT.	POINTS OF DEPARTURE & ARRIVAL		AIRCRAFT CATEGORY						GROUND TRAINER		TYPE OF PILOTING TIME					
			FROM	TO	AIRPLANE SEL		AIRPLANE MEL						DUAL RECEIVED		PILOT-IN-COMMAND			
				PAGE TOTAL														
				AMOUNT FORWARD														
				TOTAL TO DATE														

CONDITIONS OF FLIGHT										NO. INSTR. APPR.	NO. LDG. DAY / NIGHT	TOTAL DURATION OF FLIGHT		REMARKS, PROCEDURES, MANEUVERS
DAY		NIGHT		CROSS-COUNTRY		ACTUAL INSTR.		SIMULATED INSTR.						
														I certify that the statements made by me on this form are true.
														PILOT'S SIGNATURE

YEAR 20___ DATE	AIRCRAFT MAKE & MODEL	AIRCRAFT IDENT.	POINTS OF DEPARTURE & ARRIVAL		AIRCRAFT CATEGORY					GROUND TRAINER		TYPE OF PILOTING TIME					
			FROM	TO	AIRPLANE SEL		AIRPLANE MEL					DUAL RECEIVED		PILOT-IN-COMMAND			
				PAGE TOTAL													
				AMOUNT FORWARD													
				TOTAL TO DATE													

CONDITIONS OF FLIGHT										NO. INSTR. APPR.	NO. LDG. DAY / NIGHT	TOTAL DURATION OF FLIGHT		REMARKS, PROCEDURES, MANEUVERS
DAY		NIGHT		CROSS-COUNTRY		ACTUAL INSTR.		SIMULATED INSTR.						
														I certify that the statements made by me on this form are true.
														PILOT'S SIGNATURE

| YEAR 20___ DATE | AIRCRAFT MAKE & MODEL | AIRCRAFT IDENT. | POINTS OF DEPARTURE & ARRIVAL | | AIRCRAFT CATEGORY | | | | | | GROUND TRAINER | | TYPE OF PILOTING TIME | | | | | |
			FROM	TO	AIRPLANE SEL		AIRPLANE MEL						DUAL RECEIVED		PILOT-IN-COMMAND			
				PAGE TOTAL														
				AMOUNT FORWARD														
				TOTAL TO DATE														

CONDITIONS OF FLIGHT											NO. INSTR. APPR.	NO. LDG. DAY / NIGHT	TOTAL DURATION OF FLIGHT		REMARKS, PROCEDURES, MANEUVERS
DAY		NIGHT		CROSS-COUNTRY		ACTUAL INSTR.		SIMULATED INSTR.							
												/			
												/			
												/			
												/			
												/			
												/			
												/			
												/			*I certify that the statements made by me on this form are true.*
												/			
												/			PILOT'S SIGNATURE

YEAR 20___ DATE	AIRCRAFT MAKE & MODEL	AIRCRAFT IDENT.	POINTS OF DEPARTURE & ARRIVAL		AIRCRAFT CATEGORY						GROUND TRAINER		TYPE OF PILOTING TIME					
			FROM	TO	AIRPLANE SEL		AIRPLANE MEL						DUAL RECEIVED		PILOT-IN-COMMAND			
				PAGE TOTAL														
				AMOUNT FORWARD														
				TOTAL TO DATE														

CONDITIONS OF FLIGHT											NO. INSTR. APPR.	NO. LDG. DAY/NIGHT		TOTAL DURATION OF FLIGHT		REMARKS, PROCEDURES, MANEUVERS
DAY		NIGHT		CROSS-COUNTRY		ACTUAL INSTR.		SIMULATED INSTR.							
															I certify that the statements made by me on this form are true.
															PILOT'S SIGNATURE

YEAR 20___ DATE	AIRCRAFT MAKE & MODEL	AIRCRAFT IDENT.	POINTS OF DEPARTURE & ARRIVAL		AIRCRAFT CATEGORY						GROUND TRAINER	TYPE OF PILOTING TIME					
			FROM	TO	AIRPLANE SEL		AIRPLANE MEL					DUAL RECEIVED		PILOT-IN-COMMAND			
				PAGE TOTAL													
				AMOUNT FORWARD													
				TOTAL TO DATE													

CONDITIONS OF FLIGHT											NO. INSTR. APPR.	NO. LDG. DAY /NIGHT		TOTAL DURATION OF FLIGHT		REMARKS, PROCEDURES, MANEUVERS
DAY		NIGHT		CROSS-COUNTRY		ACTUAL INSTR.		SIMULATED INSTR.							
															I certify that the statements made by me on this form are true.
															PILOT'S SIGNATURE

YEAR 20___ DATE	AIRCRAFT MAKE & MODEL	AIRCRAFT IDENT.	POINTS OF DEPARTURE & ARRIVAL		AIRCRAFT CATEGORY				GROUND TRAINER	TYPE OF PILOTING TIME			
			FROM	TO	AIRPLANE SEL	AIRPLANE MEL				DUAL RECEIVED	PILOT-IN-COMMAND		
PAGE TOTAL													
AMOUNT FORWARD													
TOTAL TO DATE													

CONDITIONS OF FLIGHT											NO. INSTR. APPR.	NO. LDG. DAY/NIGHT	TOTAL DURATION OF FLIGHT		REMARKS, PROCEDURES, MANEUVERS
DAY		NIGHT		CROSS-COUNTRY		ACTUAL INSTR.		SIMULATED INSTR.							
															I certify that the statements made by me on this form are true.
															PILOT'S SIGNATURE

| YEAR 20___ DATE | AIRCRAFT MAKE & MODEL | AIRCRAFT IDENT. | POINTS OF DEPARTURE & ARRIVAL | | AIRCRAFT CATEGORY | | | | | GROUND TRAINER | | TYPE OF PILOTING TIME | | | | | |
			FROM	TO	AIRPLANE SEL		AIRPLANE MEL						DUAL RECEIVED		PILOT-IN-COMMAND			
			PAGE TOTAL															
			AMOUNT FORWARD															
			TOTAL TO DATE															

CONDITIONS OF FLIGHT												NO. INSTR. APPR.	NO. LDG. DAY / NIGHT	TOTAL DURATION OF FLIGHT		REMARKS, PROCEDURES, MANEUVERS
DAY		NIGHT		CROSS-COUNTRY		ACTUAL INSTR.		SIMULATED INSTR.								
																I certify that the statements made by me on this form are true.
																PILOT'S SIGNATURE

YEAR 20___ DATE	AIRCRAFT MAKE & MODEL	AIRCRAFT IDENT.	POINTS OF DEPARTURE & ARRIVAL		AIRCRAFT CATEGORY				GROUND TRAINER	TYPE OF PILOTING TIME			
			FROM	TO	AIRPLANE SEL	AIRPLANE MEL				DUAL RECEIVED	PILOT-IN-COMMAND		
			PAGE TOTAL										
			AMOUNT FORWARD										
			TOTAL TO DATE										

CONDITIONS OF FLIGHT											NO. INSTR. APPR.	NO. LDG. DAY / NIGHT	TOTAL DURATION OF FLIGHT		REMARKS, PROCEDURES, MANEUVERS
DAY		NIGHT		CROSS-COUNTRY		ACTUAL INSTR.		SIMULATED INSTR.							
															I certify that the statements made by me on this form are true.
															PILOT'S SIGNATURE

YEAR 20___ DATE	AIRCRAFT MAKE & MODEL	AIRCRAFT IDENT.	POINTS OF DEPARTURE & ARRIVAL		AIRCRAFT CATEGORY				GROUND TRAINER	TYPE OF PILOTING TIME					
			FROM	TO	AIRPLANE SEL		AIRPLANE MEL			DUAL RECEIVED		PILOT-IN-COMMAND			
PAGE TOTAL															
AMOUNT FORWARD															
TOTAL TO DATE															

CONDITIONS OF FLIGHT										NO. INSTR. APPR.	NO. LDG. DAY / NIGHT	TOTAL DURATION OF FLIGHT		REMARKS, PROCEDURES, MANEUVERS
DAY		NIGHT		CROSS-COUNTRY		ACTUAL INSTR.		SIMULATED INSTR.						
														I certify that the statements made by me on this form are true.
														PILOT'S SIGNATURE

YEAR 20___ DATE	AIRCRAFT MAKE & MODEL	AIRCRAFT IDENT.	POINTS OF DEPARTURE & ARRIVAL		AIRCRAFT CATEGORY				GROUND TRAINER		TYPE OF PILOTING TIME			
			FROM	TO	AIRPLANE SEL		AIRPLANE MEL				DUAL RECEIVED		PILOT-IN-COMMAND	
PAGE TOTAL														
AMOUNT FORWARD														
TOTAL TO DATE														

CONDITIONS OF FLIGHT													NO. INSTR. APPR.	NO. LDG. DAY / NIGHT	TOTAL DURATION OF FLIGHT		REMARKS, PROCEDURES, MANEUVERS
DAY		NIGHT		CROSS-COUNTRY		ACTUAL INSTR.		SIMULATED INSTR.									
																	I certify that the statements made by me on this form are true.
																	PILOT'S SIGNATURE

YEAR 20___ DATE	AIRCRAFT MAKE & MODEL	AIRCRAFT IDENT.	POINTS OF DEPARTURE & ARRIVAL		AIRCRAFT CATEGORY					GROUND TRAINER	TYPE OF PILOTING TIME			
			FROM	TO	AIRPLANE SEL		AIRPLANE MEL				DUAL RECEIVED		PILOT-IN-COMMAND	
PAGE TOTAL														
AMOUNT FORWARD														
TOTAL TO DATE														

CONDITIONS OF FLIGHT										NO. INSTR. APPR.	NO. LDG. DAY / NIGHT		TOTAL DURATION OF FLIGHT		REMARKS, PROCEDURES, MANEUVERS
DAY		NIGHT		CROSS-COUNTRY		ACTUAL INSTR.		SIMULATED INSTR.							
															I certify that the statements made by me on this form are true.
															PILOT'S SIGNATURE

YEAR 20___ DATE	AIRCRAFT MAKE & MODEL	AIRCRAFT IDENT.	POINTS OF DEPARTURE & ARRIVAL		AIRCRAFT CATEGORY				GROUND TRAINER	TYPE OF PILOTING TIME			
			FROM	TO	AIRPLANE SEL		AIRPLANE MEL			DUAL RECEIVED		PILOT-IN-COMMAND	
PAGE TOTAL													
AMOUNT FORWARD													
TOTAL TO DATE													

CONDITIONS OF FLIGHT											NO. INSTR. APPR.	NO. LDG. DAY / NIGHT	TOTAL DURATION OF FLIGHT		REMARKS, PROCEDURES, MANEUVERS
DAY		NIGHT		CROSS-COUNTRY		ACTUAL INSTR.		SIMULATED INSTR.							
												/			
												/			
												/			
												/			
												/			
												/			
												/			
												/			*I certify that the statements made by me on this form are true.*
												/			
												/			PILOT'S SIGNATURE

YEAR 20___ DATE	AIRCRAFT MAKE & MODEL	AIRCRAFT IDENT.	POINTS OF DEPARTURE & ARRIVAL		AIRCRAFT CATEGORY				GROUND TRAINER		TYPE OF PILOTING TIME				
			FROM	TO	AIRPLANE SEL		AIRPLANE MEL				DUAL RECEIVED		PILOT-IN-COMMAND		
PAGE TOTAL															
AMOUNT FORWARD															
TOTAL TO DATE															

CONDITIONS OF FLIGHT											NO. INSTR. APPR.	NO. LDG. DAY / NIGHT	TOTAL DURATION OF FLIGHT		REMARKS, PROCEDURES, MANEUVERS
DAY		NIGHT		CROSS-COUNTRY		ACTUAL INSTR.		SIMULATED INSTR.							
															I certify that the statements made by me on this form are true.
															PILOT'S SIGNATURE

YEAR 20___ DATE	AIRCRAFT MAKE & MODEL	AIRCRAFT IDENT.	POINTS OF DEPARTURE & ARRIVAL		AIRCRAFT CATEGORY						GROUND TRAINER	TYPE OF PILOTING TIME					
			FROM	TO	AIRPLANE SEL		AIRPLANE MEL					DUAL RECEIVED		PILOT-IN-COMMAND			
				PAGE TOTAL													
				AMOUNT FORWARD													
				TOTAL TO DATE													

CONDITIONS OF FLIGHT											NO. INSTR. APPR.	NO. LDG. DAY / NIGHT	TOTAL DURATION OF FLIGHT		REMARKS, PROCEDURES, MANEUVERS
DAY		NIGHT		CROSS-COUNTRY		ACTUAL INSTR.		SIMULATED INSTR.							
															I certify that the statements made by me on this form are true.
															PILOT'S SIGNATURE

| YEAR 20__ DATE | AIRCRAFT MAKE & MODEL | AIRCRAFT IDENT. | POINTS OF DEPARTURE & ARRIVAL | | AIRCRAFT CATEGORY | | | | GROUND TRAINER | TYPE OF PILOTING TIME | | | |
			FROM	TO	AIRPLANE SEL		AIRPLANE MEL			DUAL RECEIVED		PILOT-IN-COMMAND	
PAGE TOTAL													
AMOUNT FORWARD													
TOTAL TO DATE													

CONDITIONS OF FLIGHT												NO. INSTR. APPR.	NO. LDG. DAY / NIGHT		TOTAL DURATION OF FLIGHT		REMARKS, PROCEDURES, MANEUVERS
DAY		NIGHT		CROSS-COUNTRY		ACTUAL INSTR.		SIMULATED INSTR.									
																	I certify that the statements made by me on this form are true.
																	PILOT'S SIGNATURE

YEAR 20___ DATE	AIRCRAFT MAKE & MODEL	AIRCRAFT IDENT.	POINTS OF DEPARTURE & ARRIVAL		AIRCRAFT CATEGORY				GROUND TRAINER	TYPE OF PILOTING TIME			
			FROM	TO	AIRPLANE SEL	AIRPLANE MEL				DUAL RECEIVED	PILOT-IN-COMMAND		
			PAGE TOTAL										
			AMOUNT FORWARD										
			TOTAL TO DATE										

CONDITIONS OF FLIGHT										NO. INSTR. APPR.	NO. LDG. DAY / NIGHT	TOTAL DURATION OF FLIGHT		REMARKS, PROCEDURES, MANEUVERS
DAY		NIGHT		CROSS-COUNTRY		ACTUAL INSTR.		SIMULATED INSTR.						
														I certify that the statements made by me on this form are true.
														PILOT'S SIGNATURE

YEAR 20___ DATE	AIRCRAFT MAKE & MODEL	AIRCRAFT IDENT.	POINTS OF DEPARTURE & ARRIVAL		AIRCRAFT CATEGORY						GROUND TRAINER		TYPE OF PILOTING TIME				
			FROM	TO	AIRPLANE SEL		AIRPLANE MEL						DUAL RECEIVED		PILOT-IN-COMMAND		
			PAGE TOTAL														
			AMOUNT FORWARD														
			TOTAL TO DATE														

CONDITIONS OF FLIGHT										NO. INSTR. APPR.	NO. LDG.		TOTAL DURATION OF FLIGHT		REMARKS, PROCEDURES, MANEUVERS
DAY		NIGHT		CROSS-COUNTRY		ACTUAL INSTR.		SIMULATED INSTR.			DAY / NIGHT				
															I certify that the statements made by me on this form are true.
															PILOT'S SIGNATURE

YEAR 20___ DATE	AIRCRAFT MAKE & MODEL	AIRCRAFT IDENT.	POINTS OF DEPARTURE & ARRIVAL		AIRCRAFT CATEGORY					GROUND TRAINER		TYPE OF PILOTING TIME				
			FROM	TO	AIRPLANE SEL		AIRPLANE MEL					DUAL RECEIVED		PILOT-IN-COMMAND		
			PAGE TOTAL													
			AMOUNT FORWARD													
			TOTAL TO DATE													

CONDITIONS OF FLIGHT										NO. INSTR. APPR.	NO. LDG. DAY / NIGHT		TOTAL DURATION OF FLIGHT		REMARKS, PROCEDURES, MANEUVERS
DAY		NIGHT		CROSS-COUNTRY		ACTUAL INSTR.		SIMULATED INSTR.							
															I certify that the statements made by me on this form are true.
															PILOT'S SIGNATURE

YEAR 20___ DATE	AIRCRAFT MAKE & MODEL	AIRCRAFT IDENT.	POINTS OF DEPARTURE & ARRIVAL		AIRCRAFT CATEGORY					GROUND TRAINER	TYPE OF PILOTING TIME					
			FROM	TO	AIRPLANE SEL		AIRPLANE MEL					DUAL RECEIVED		PILOT-IN-COMMAND		
PAGE TOTAL																
AMOUNT FORWARD																
TOTAL TO DATE																

CONDITIONS OF FLIGHT											NO. INSTR. APPR.	NO. LDG. DAY/NIGHT	TOTAL DURATION OF FLIGHT		REMARKS, PROCEDURES, MANEUVERS
DAY		NIGHT		CROSS-COUNTRY		ACTUAL INSTR.		SIMULATED INSTR.							
															I certify that the statements made by me on this form are true.
															PILOT'S SIGNATURE

YEAR 20___ DATE	AIRCRAFT MAKE & MODEL	AIRCRAFT IDENT.	POINTS OF DEPARTURE & ARRIVAL		AIRCRAFT CATEGORY			GROUND TRAINER	TYPE OF PILOTING TIME		
			FROM	TO	AIRPLANE SEL	AIRPLANE MEL			DUAL RECEIVED	PILOT-IN-COMMAND	
			PAGE TOTAL								
			AMOUNT FORWARD								
			TOTAL TO DATE								

CONDITIONS OF FLIGHT										NO. INSTR. APPR.	NO. LDG. DAY/NIGHT	TOTAL DURATION OF FLIGHT		REMARKS, PROCEDURES, MANEUVERS
DAY		NIGHT		CROSS-COUNTRY		ACTUAL INSTR.		SIMULATED INSTR.						
														I certify that the statements made by me on this form are true.
														PILOT'S SIGNATURE

YEAR 20___ DATE	AIRCRAFT MAKE & MODEL	AIRCRAFT IDENT.	POINTS OF DEPARTURE & ARRIVAL		AIRCRAFT CATEGORY						GROUND TRAINER		TYPE OF PILOTING TIME					
			FROM	TO	AIRPLANE SEL		AIRPLANE MEL						DUAL RECEIVED		PILOT-IN-COMMAND			
PAGE TOTAL																		
AMOUNT FORWARD																		
TOTAL TO DATE																		

CONDITIONS OF FLIGHT												NO. INSTR. APPR.	NO. LDG. DAY / NIGHT	TOTAL DURATION OF FLIGHT		REMARKS, PROCEDURES, MANEUVERS
DAY		NIGHT		CROSS-COUNTRY		ACTUAL INSTR.		SIMULATED INSTR.								
																I certify that the statements made by me on this form are true.
																PILOT'S SIGNATURE

YEAR 20___ DATE	AIRCRAFT MAKE & MODEL	AIRCRAFT IDENT.	POINTS OF DEPARTURE & ARRIVAL		AIRCRAFT CATEGORY					GROUND TRAINER		TYPE OF PILOTING TIME					
			FROM	TO	AIRPLANE SEL		AIRPLANE MEL					DUAL RECEIVED		PILOT-IN-COMMAND			
			PAGE TOTAL														
			AMOUNT FORWARD														
			TOTAL TO DATE														

CONDITIONS OF FLIGHT										NO. INSTR. APPR.	NO. LDG. DAY / NIGHT	TOTAL DURATION OF FLIGHT		REMARKS, PROCEDURES, MANEUVERS
DAY		NIGHT		CROSS-COUNTRY		ACTUAL INSTR.		SIMULATED INSTR.						
														I certify that the statements made by me on this form are true.
														PILOT'S SIGNATURE

YEAR 20___ DATE	AIRCRAFT MAKE & MODEL	AIRCRAFT IDENT.	POINTS OF DEPARTURE & ARRIVAL		AIRCRAFT CATEGORY					GROUND TRAINER	TYPE OF PILOTING TIME			
			FROM	TO	AIRPLANE SEL		AIRPLANE MEL				DUAL RECEIVED		PILOT-IN-COMMAND	
PAGE TOTAL														
AMOUNT FORWARD														
TOTAL TO DATE														

CONDITIONS OF FLIGHT													NO. INSTR. APPR.	NO. LDG. DAY/NIGHT	TOTAL DURATION OF FLIGHT		REMARKS, PROCEDURES, MANEUVERS
DAY		NIGHT		CROSS-COUNTRY		ACTUAL INSTR.		SIMULATED INSTR.									
																	I certify that the statements made by me on this form are true.
																	PILOT'S SIGNATURE

YEAR 20___ DATE	AIRCRAFT MAKE & MODEL	AIRCRAFT IDENT.	POINTS OF DEPARTURE & ARRIVAL		AIRCRAFT CATEGORY			GROUND TRAINER	TYPE OF PILOTING TIME			
			FROM	TO	AIRPLANE SEL	AIRPLANE MEL			DUAL RECEIVED	PILOT-IN-COMMAND		
			PAGE TOTAL									
			AMOUNT FORWARD									
			TOTAL TO DATE									

CONDITIONS OF FLIGHT										NO. INSTR. APPR.	NO. LDG. DAY / NIGHT	TOTAL DURATION OF FLIGHT		REMARKS, PROCEDURES, MANEUVERS
DAY		NIGHT		CROSS-COUNTRY		ACTUAL INSTR.		SIMULATED INSTR.						
														I certify that the statements made by me on this form are true.
														PILOT'S SIGNATURE

YEAR 20___ DATE	AIRCRAFT MAKE & MODEL	AIRCRAFT IDENT.	POINTS OF DEPARTURE & ARRIVAL		AIRCRAFT CATEGORY						GROUND TRAINER	TYPE OF PILOTING TIME					
			FROM	TO	AIRPLANE SEL		AIRPLANE MEL					DUAL RECEIVED		PILOT-IN-COMMAND			
				PAGE TOTAL													
				AMOUNT FORWARD													
				TOTAL TO DATE													

CONDITIONS OF FLIGHT												NO. INSTR. APPR.	NO. LDG. DAY/NIGHT	TOTAL DURATION OF FLIGHT		REMARKS, PROCEDURES, MANEUVERS
DAY		NIGHT		CROSS-COUNTRY		ACTUAL INSTR.		SIMULATED INSTR.								
																I certify that the statements made by me on this form are true.
																PILOT'S SIGNATURE

YEAR 20___ DATE	AIRCRAFT MAKE & MODEL	AIRCRAFT IDENT.	POINTS OF DEPARTURE & ARRIVAL		AIRCRAFT CATEGORY			GROUND TRAINER	TYPE OF PILOTING TIME		
			FROM	TO	AIRPLANE SEL	AIRPLANE MEL			DUAL RECEIVED	PILOT-IN-COMMAND	
			PAGE TOTAL								
			AMOUNT FORWARD								
			TOTAL TO DATE								

CONDITIONS OF FLIGHT										NO. INSTR. APPR.	NO. LDG. DAY / NIGHT	TOTAL DURATION OF FLIGHT		REMARKS, PROCEDURES, MANEUVERS
DAY		NIGHT		CROSS-COUNTRY		ACTUAL INSTR.		SIMULATED INSTR.						
														I certify that the statements made by me on this form are true.
														PILOT'S SIGNATURE

YEAR 20___ DATE	AIRCRAFT MAKE & MODEL	AIRCRAFT IDENT.	POINTS OF DEPARTURE & ARRIVAL		AIRCRAFT CATEGORY				GROUND TRAINER	TYPE OF PILOTING TIME			
			FROM	TO	AIRPLANE SEL	AIRPLANE MEL				DUAL RECEIVED	PILOT-IN-COMMAND		
PAGE TOTAL													
AMOUNT FORWARD													
TOTAL TO DATE													

CONDITIONS OF FLIGHT										NO. INSTR. APPR.	NO. LDG. DAY/NIGHT	TOTAL DURATION OF FLIGHT		REMARKS, PROCEDURES, MANEUVERS
DAY		NIGHT		CROSS-COUNTRY		ACTUAL INSTR.		SIMULATED INSTR.						
														I certify that the statements made by me on this form are true.
														PILOT'S SIGNATURE

YEAR 20___ DATE	AIRCRAFT MAKE & MODEL	AIRCRAFT IDENT.	POINTS OF DEPARTURE & ARRIVAL		AIRCRAFT CATEGORY						GROUND TRAINER		TYPE OF PILOTING TIME					
			FROM	TO	AIRPLANE SEL		AIRPLANE MEL						DUAL RECEIVED		PILOT-IN-COMMAND			
				PAGE TOTAL														
				AMOUNT FORWARD														
				TOTAL TO DATE														

CONDITIONS OF FLIGHT										NO. INSTR. APPR.	NO. LDG. DAY / NIGHT	TOTAL DURATION OF FLIGHT		REMARKS, PROCEDURES, MANEUVERS
DAY		NIGHT		CROSS-COUNTRY		ACTUAL INSTR.		SIMULATED INSTR.						
														I certify that the statements made by me on this form are true.
														PILOT'S SIGNATURE

YEAR 20__ DATE	AIRCRAFT MAKE & MODEL	AIRCRAFT IDENT.	POINTS OF DEPARTURE & ARRIVAL		AIRCRAFT CATEGORY				GROUND TRAINER	TYPE OF PILOTING TIME			
			FROM	TO	AIRPLANE SEL	AIRPLANE MEL				DUAL RECEIVED	PILOT-IN-COMMAND		
			PAGE TOTAL										
			AMOUNT FORWARD										
			TOTAL TO DATE										

CONDITIONS OF FLIGHT										NO. INSTR. APPR.	NO. LDG. DAY / NIGHT	TOTAL DURATION OF FLIGHT		REMARKS, PROCEDURES, MANEUVERS
DAY		NIGHT		CROSS-COUNTRY		ACTUAL INSTR.		SIMULATED INSTR.						
														I certify that the statements made by me on this form are true.
														PILOT'S SIGNATURE

| YEAR 20___ DATE | AIRCRAFT MAKE & MODEL | AIRCRAFT IDENT. | POINTS OF DEPARTURE & ARRIVAL | | AIRCRAFT CATEGORY | | | | GROUND TRAINER | TYPE OF PILOTING TIME | | | |
			FROM	TO	AIRPLANE SEL	AIRPLANE MEL				DUAL RECEIVED	PILOT-IN-COMMAND		
PAGE TOTAL													
AMOUNT FORWARD													
TOTAL TO DATE													

CONDITIONS OF FLIGHT												NO. INSTR. APPR.	NO. LDG. DAY / NIGHT	TOTAL DURATION OF FLIGHT		REMARKS, PROCEDURES, MANEUVERS
DAY		NIGHT		CROSS-COUNTRY		ACTUAL INSTR.		SIMULATED INSTR.								
																I certify that the statements made by me on this form are true.
																PILOT'S SIGNATURE

| YEAR 20___ DATE | AIRCRAFT MAKE & MODEL | AIRCRAFT IDENT. | POINTS OF DEPARTURE & ARRIVAL | | AIRCRAFT CATEGORY | | | | GROUND TRAINER | | TYPE OF PILOTING TIME | | | |
			FROM	TO	AIRPLANE SEL		AIRPLANE MEL				DUAL RECEIVED		PILOT-IN-COMMAND	
				PAGE TOTAL										
				AMOUNT FORWARD										
				TOTAL TO DATE										

CONDITIONS OF FLIGHT											NO. INSTR. APPR.	NO. LDG. DAY / NIGHT		TOTAL DURATION OF FLIGHT		REMARKS, PROCEDURES, MANEUVERS
DAY		NIGHT		CROSS-COUNTRY		ACTUAL INSTR.		SIMULATED INSTR.								
																I certify that the statements made by me on this form are true.
																PILOT'S SIGNATURE

YEAR 20___ DATE	AIRCRAFT MAKE & MODEL	AIRCRAFT IDENT.	POINTS OF DEPARTURE & ARRIVAL		AIRCRAFT CATEGORY				GROUND TRAINER	TYPE OF PILOTING TIME			
			FROM	TO	AIRPLANE SEL	AIRPLANE MEL				DUAL RECEIVED	PILOT-IN-COMMAND		
			PAGE TOTAL										
			AMOUNT FORWARD										
			TOTAL TO DATE										

CONDITIONS OF FLIGHT										NO. INSTR. APPR.	NO. LDG. DAY / NIGHT		TOTAL DURATION OF FLIGHT		REMARKS, PROCEDURES, MANEUVERS
DAY		NIGHT		CROSS-COUNTRY		ACTUAL INSTR.		SIMULATED INSTR.							
															I certify that the statements made by me on this form are true.
															PILOT'S SIGNATURE

YEAR 20___ DATE	AIRCRAFT MAKE & MODEL	AIRCRAFT IDENT.	POINTS OF DEPARTURE & ARRIVAL		AIRCRAFT CATEGORY				GROUND TRAINER	TYPE OF PILOTING TIME		
			FROM	TO	AIRPLANE SEL	AIRPLANE MEL				DUAL RECEIVED	PILOT-IN-COMMAND	
				PAGE TOTAL								
				AMOUNT FORWARD								
				TOTAL TO DATE								

CONDITIONS OF FLIGHT										NO. INSTR. APPR.	NO. LDG. DAY / NIGHT	TOTAL DURATION OF FLIGHT		REMARKS, PROCEDURES, MANEUVERS
DAY		NIGHT		CROSS-COUNTRY		ACTUAL INSTR.		SIMULATED INSTR.						
														I certify that the statements made by me on this form are true.
														PILOT'S SIGNATURE

YEAR 20___ DATE	AIRCRAFT MAKE & MODEL	AIRCRAFT IDENT.	POINTS OF DEPARTURE & ARRIVAL		AIRCRAFT CATEGORY				GROUND TRAINER	TYPE OF PILOTING TIME			
			FROM	TO	AIRPLANE SEL	AIRPLANE MEL				DUAL RECEIVED	PILOT-IN-COMMAND		
			PAGE TOTAL										
			AMOUNT FORWARD										
			TOTAL TO DATE										

CONDITIONS OF FLIGHT												NO. INSTR. APPR.	NO. LDG. DAY / NIGHT	TOTAL DURATION OF FLIGHT		REMARKS, PROCEDURES, MANEUVERS
DAY		NIGHT		CROSS-COUNTRY		ACTUAL INSTR.		SIMULATED INSTR.								
																I certify that the statements made by me on this form are true.
																PILOT'S SIGNATURE

YEAR 20___ DATE	AIRCRAFT MAKE & MODEL	AIRCRAFT IDENT.	POINTS OF DEPARTURE & ARRIVAL		AIRCRAFT CATEGORY				GROUND TRAINER	TYPE OF PILOTING TIME			
			FROM	TO	AIRPLANE SEL		AIRPLANE MEL			DUAL RECEIVED		PILOT-IN-COMMAND	
				PAGE TOTAL									
				AMOUNT FORWARD									
				TOTAL TO DATE									

CONDITIONS OF FLIGHT										NO. INSTR. APPR.	NO. LDG. DAY/NIGHT		TOTAL DURATION OF FLIGHT		REMARKS, PROCEDURES, MANEUVERS
DAY		NIGHT		CROSS-COUNTRY		ACTUAL INSTR.		SIMULATED INSTR.							
															I certify that the statements made by me on this form are true.
															PILOT'S SIGNATURE

| YEAR 20___ DATE | AIRCRAFT MAKE & MODEL | AIRCRAFT IDENT. | POINTS OF DEPARTURE & ARRIVAL | | AIRCRAFT CATEGORY | | | | GROUND TRAINER | | TYPE OF PILOTING TIME | | | |
			FROM	TO	AIRPLANE SEL	AIRPLANE MEL					DUAL RECEIVED	PILOT-IN-COMMAND		
			PAGE TOTAL											
			AMOUNT FORWARD											
			TOTAL TO DATE											

CONDITIONS OF FLIGHT										NO. INSTR. APPR.	NO. LDG.		TOTAL DURATION OF FLIGHT		REMARKS, PROCEDURES, MANEUVERS
DAY		NIGHT		CROSS-COUNTRY		ACTUAL INSTR.		SIMULATED INSTR.			DAY / NIGHT				
															I certify that the statements made by me on this form are true.
															PILOT'S SIGNATURE

YEAR 20___ DATE	AIRCRAFT MAKE & MODEL	AIRCRAFT IDENT.	POINTS OF DEPARTURE & ARRIVAL		AIRCRAFT CATEGORY						GROUND TRAINER		TYPE OF PILOTING TIME					
			FROM	TO	AIRPLANE SEL		AIRPLANE MEL						DUAL RECEIVED		PILOT-IN-COMMAND			
				PAGE TOTAL														
				AMOUNT FORWARD														
				TOTAL TO DATE														

CONDITIONS OF FLIGHT										NO. INSTR. APPR.	NO. LDG. DAY / NIGHT		TOTAL DURATION OF FLIGHT		REMARKS, PROCEDURES, MANEUVERS
DAY		NIGHT		CROSS-COUNTRY		ACTUAL INSTR.		SIMULATED INSTR.							
															I certify that the statements made by me on this form are true.
															PILOT'S SIGNATURE

| YEAR 20___ DATE | AIRCRAFT MAKE & MODEL | AIRCRAFT IDENT. | POINTS OF DEPARTURE & ARRIVAL | | AIRCRAFT CATEGORY | | | | | GROUND TRAINER | | TYPE OF PILOTING TIME | | | | | |
			FROM	TO	AIRPLANE SEL		AIRPLANE MEL					DUAL RECEIVED		PILOT-IN-COMMAND			
				PAGE TOTAL													
				AMOUNT FORWARD													
				TOTAL TO DATE													

CONDITIONS OF FLIGHT											NO. INSTR. APPR.	NO. LDG. DAY/NIGHT	TOTAL DURATION OF FLIGHT	REMARKS, PROCEDURES, MANEUVERS
DAY		NIGHT		CROSS-COUNTRY		ACTUAL INSTR.		SIMULATED INSTR.						
														I certify that the statements made by me on this form are true.
														PILOT'S SIGNATURE

YEAR 20___ DATE	AIRCRAFT MAKE & MODEL	AIRCRAFT IDENT.	POINTS OF DEPARTURE & ARRIVAL		AIRCRAFT CATEGORY			GROUND TRAINER	TYPE OF PILOTING TIME			
			FROM	TO	AIRPLANE SEL	AIRPLANE MEL			DUAL RECEIVED	PILOT-IN-COMMAND		
PAGE TOTAL												
AMOUNT FORWARD												
TOTAL TO DATE												

CONDITIONS OF FLIGHT										NO. INSTR. APPR.	NO. LDG. DAY / NIGHT		TOTAL DURATION OF FLIGHT		REMARKS, PROCEDURES, MANEUVERS
DAY		NIGHT		CROSS-COUNTRY		ACTUAL INSTR.		SIMULATED INSTR.							
															I certify that the statements made by me on this form are true.
															PILOT'S SIGNATURE

YEAR 20___ DATE	AIRCRAFT MAKE & MODEL	AIRCRAFT IDENT.	POINTS OF DEPARTURE & ARRIVAL		AIRCRAFT CATEGORY						GROUND TRAINER		TYPE OF PILOTING TIME					
			FROM	TO	AIRPLANE SEL		AIRPLANE MEL						DUAL RECEIVED		PILOT-IN-COMMAND			
				PAGE TOTAL														
				AMOUNT FORWARD														
				TOTAL TO DATE														

CONDITIONS OF FLIGHT											NO. INSTR. APPR.	NO. LDG. DAY / NIGHT	TOTAL DURATION OF FLIGHT		REMARKS, PROCEDURES, MANEUVERS
DAY		NIGHT		CROSS-COUNTRY		ACTUAL INSTR.		SIMULATED INSTR.							
															I certify that the statements made by me on this form are true.
															PILOT'S SIGNATURE

| YEAR 20___ DATE | AIRCRAFT MAKE & MODEL | AIRCRAFT IDENT. | POINTS OF DEPARTURE & ARRIVAL | | AIRCRAFT CATEGORY | | | | | | GROUND TRAINER | | TYPE OF PILOTING TIME | | | | | |
			FROM	TO	AIRPLANE SEL		AIRPLANE MEL						DUAL RECEIVED		PILOT-IN-COMMAND			
				PAGE TOTAL														
				AMOUNT FORWARD														
				TOTAL TO DATE														

CONDITIONS OF FLIGHT											NO. INSTR. APPR.	NO. LDG. DAY / NIGHT		TOTAL DURATION OF FLIGHT		REMARKS, PROCEDURES, MANEUVERS
DAY		NIGHT		CROSS-COUNTRY		ACTUAL INSTR.		SIMULATED INSTR.								
																I certify that the statements made by me on this form are true.
																PILOT'S SIGNATURE

YEAR 20___ DATE	AIRCRAFT MAKE & MODEL	AIRCRAFT IDENT.	POINTS OF DEPARTURE & ARRIVAL		AIRCRAFT CATEGORY				GROUND TRAINER		TYPE OF PILOTING TIME			
			FROM	TO	AIRPLANE SEL	AIRPLANE MEL					DUAL RECEIVED	PILOT-IN-COMMAND		
PAGE TOTAL														
AMOUNT FORWARD														
TOTAL TO DATE														

CONDITIONS OF FLIGHT											NO. INSTR. APPR.	NO. LDG. DAY /NIGHT	TOTAL DURATION OF FLIGHT		REMARKS, PROCEDURES, MANEUVERS
DAY		NIGHT		CROSS-COUNTRY		ACTUAL INSTR.		SIMULATED INSTR.							
															I certify that the statements made by me on this form are true.
															PILOT'S SIGNATURE

YEAR 20___ DATE	AIRCRAFT MAKE & MODEL	AIRCRAFT IDENT.	POINTS OF DEPARTURE & ARRIVAL		AIRCRAFT CATEGORY				GROUND TRAINER	TYPE OF PILOTING TIME			
			FROM	TO	AIRPLANE SEL	AIRPLANE MEL				DUAL RECEIVED	PILOT-IN-COMMAND		
PAGE TOTAL													
AMOUNT FORWARD													
TOTAL TO DATE													

CONDITIONS OF FLIGHT												NO. INSTR. APPR.	NO. LDG. DAY/NIGHT	TOTAL DURATION OF FLIGHT		REMARKS, PROCEDURES, MANEUVERS
DAY		NIGHT		CROSS-COUNTRY		ACTUAL INSTR.		SIMULATED INSTR.								
																I certify that the statements made by me on this form are true.
																PILOT'S SIGNATURE

| YEAR 20__ DATE | AIRCRAFT MAKE & MODEL | AIRCRAFT IDENT. | POINTS OF DEPARTURE & ARRIVAL | | AIRCRAFT CATEGORY | | | | | GROUND TRAINER | | TYPE OF PILOTING TIME | | | | | |
			FROM	TO	AIRPLANE SEL		AIRPLANE MEL					DUAL RECEIVED		PILOT-IN-COMMAND			
			PAGE TOTAL														
			AMOUNT FORWARD														
			TOTAL TO DATE														

CONDITIONS OF FLIGHT										NO. INSTR. APPR.	NO. LDG. DAY / NIGHT		TOTAL DURATION OF FLIGHT		REMARKS, PROCEDURES, MANEUVERS
DAY		NIGHT		CROSS-COUNTRY		ACTUAL INSTR.		SIMULATED INSTR.							
															I certify that the statements made by me on this form are true.
															PILOT'S SIGNATURE

YEAR 20___ DATE	AIRCRAFT MAKE & MODEL	AIRCRAFT IDENT.	POINTS OF DEPARTURE & ARRIVAL		AIRCRAFT CATEGORY					GROUND TRAINER		TYPE OF PILOTING TIME					
			FROM	TO	AIRPLANE SEL		AIRPLANE MEL					DUAL RECEIVED		PILOT-IN-COMMAND			
				PAGE TOTAL													
				AMOUNT FORWARD													
				TOTAL TO DATE													

CONDITIONS OF FLIGHT											NO. INSTR. APPR.	NO. LDG. DAY / NIGHT	TOTAL DURATION OF FLIGHT		REMARKS, PROCEDURES, MANEUVERS
DAY		NIGHT		CROSS-COUNTRY		ACTUAL INSTR.		SIMULATED INSTR.							
															I certify that the statements made by me on this form are true.
															PILOT'S SIGNATURE

YEAR 20___ DATE	AIRCRAFT MAKE & MODEL	AIRCRAFT IDENT.	POINTS OF DEPARTURE & ARRIVAL		AIRCRAFT CATEGORY						GROUND TRAINER	TYPE OF PILOTING TIME					
			FROM	TO	AIRPLANE SEL		AIRPLANE MEL					DUAL RECEIVED		PILOT-IN-COMMAND			
				PAGE TOTAL													
				AMOUNT FORWARD													
				TOTAL TO DATE													

CONDITIONS OF FLIGHT										NO. INSTR. APPR.	NO. LDG. DAY / NIGHT		TOTAL DURATION OF FLIGHT		REMARKS, PROCEDURES, MANEUVERS
DAY		NIGHT		CROSS-COUNTRY		ACTUAL INSTR.		SIMULATED INSTR.							
															I certify that the statements made by me on this form are true.
															PILOT'S SIGNATURE

| YEAR 20__ DATE | AIRCRAFT MAKE & MODEL | AIRCRAFT IDENT. | POINTS OF DEPARTURE & ARRIVAL | | AIRCRAFT CATEGORY | | | | | | GROUND TRAINER | | TYPE OF PILOTING TIME | | | | | |
			FROM	TO	AIRPLANE SEL		AIRPLANE MEL						DUAL RECEIVED		PILOT-IN-COMMAND			
				PAGE TOTAL														
				AMOUNT FORWARD														
				TOTAL TO DATE														

CONDITIONS OF FLIGHT												NO. INSTR. APPR.	NO. LDG. DAY / NIGHT	TOTAL DURATION OF FLIGHT		REMARKS, PROCEDURES, MANEUVERS
DAY		NIGHT		CROSS-COUNTRY		ACTUAL INSTR.		SIMULATED INSTR.								
																I certify that the statements made by me on this form are true.
																PILOT'S SIGNATURE

YEAR 20___ DATE	AIRCRAFT MAKE & MODEL	AIRCRAFT IDENT.	POINTS OF DEPARTURE & ARRIVAL		AIRCRAFT CATEGORY					GROUND TRAINER	TYPE OF PILOTING TIME					
			FROM	TO	AIRPLANE SEL		AIRPLANE MEL					DUAL RECEIVED		PILOT-IN-COMMAND		
			PAGE TOTAL													
			AMOUNT FORWARD													
			TOTAL TO DATE													

CONDITIONS OF FLIGHT											NO. INSTR. APPR.	NO. LDG. DAY/NIGHT		TOTAL DURATION OF FLIGHT		REMARKS, PROCEDURES, MANEUVERS
DAY		NIGHT		CROSS-COUNTRY		ACTUAL INSTR.		SIMULATED INSTR.								
																I certify that the statements made by me on this form are true.
																PILOT'S SIGNATURE

YEAR 20___ DATE	AIRCRAFT MAKE & MODEL	AIRCRAFT IDENT.	POINTS OF DEPARTURE & ARRIVAL		AIRCRAFT CATEGORY						GROUND TRAINER		TYPE OF PILOTING TIME					
			FROM	TO	AIRPLANE SEL		AIRPLANE MEL						DUAL RECEIVED		PILOT-IN-COMMAND			
				PAGE TOTAL														
				AMOUNT FORWARD														
				TOTAL TO DATE														

CONDITIONS OF FLIGHT										NO. INSTR. APPR.	NO. LDG. DAY / NIGHT		TOTAL DURATION OF FLIGHT		REMARKS, PROCEDURES, MANEUVERS
DAY		NIGHT		CROSS-COUNTRY		ACTUAL INSTR.		SIMULATED INSTR.							
															I certify that the statements made by me on this form are true.
															PILOT'S SIGNATURE

YEAR 20___ DATE	AIRCRAFT MAKE & MODEL	AIRCRAFT IDENT.	POINTS OF DEPARTURE & ARRIVAL		AIRCRAFT CATEGORY				GROUND TRAINER	TYPE OF PILOTING TIME			
			FROM	TO	AIRPLANE SEL	AIRPLANE MEL				DUAL RECEIVED	PILOT-IN-COMMAND		
			PAGE TOTAL										
			AMOUNT FORWARD										
			TOTAL TO DATE										

CONDITIONS OF FLIGHT										NO. INSTR. APPR.	NO. LDG. DAY/NIGHT	TOTAL DURATION OF FLIGHT	REMARKS, PROCEDURES, MANEUVERS
DAY		NIGHT		CROSS-COUNTRY		ACTUAL INSTR.		SIMULATED INSTR.					
													I certify that the statements made by me on this form are true.
													PILOT'S SIGNATURE

| YEAR 20__ DATE | AIRCRAFT MAKE & MODEL | AIRCRAFT IDENT. | POINTS OF DEPARTURE & ARRIVAL | | AIRCRAFT CATEGORY | | | | | GROUND TRAINER | | TYPE OF PILOTING TIME | | | | | |
			FROM	TO	AIRPLANE SEL		AIRPLANE MEL						DUAL RECEIVED		PILOT-IN-COMMAND		
				PAGE TOTAL													
				AMOUNT FORWARD													
				TOTAL TO DATE													

CONDITIONS OF FLIGHT											NO. INSTR. APPR.	NO. LDG. DAY/NIGHT		TOTAL DURATION OF FLIGHT		REMARKS, PROCEDURES, MANEUVERS
DAY		NIGHT		CROSS-COUNTRY		ACTUAL INSTR.		SIMULATED INSTR.							
															I certify that the statements made by me on this form are true.
															PILOT'S SIGNATURE

YEAR 20___ DATE	AIRCRAFT MAKE & MODEL	AIRCRAFT IDENT.	POINTS OF DEPARTURE & ARRIVAL		AIRCRAFT CATEGORY				GROUND TRAINER	TYPE OF PILOTING TIME			
			FROM	TO	AIRPLANE SEL	AIRPLANE MEL				DUAL RECEIVED	PILOT-IN-COMMAND		
			PAGE TOTAL										
			AMOUNT FORWARD										
			TOTAL TO DATE										

CONDITIONS OF FLIGHT										NO. INSTR. APPR.	NO. LDG. DAY / NIGHT	TOTAL DURATION OF FLIGHT		REMARKS, PROCEDURES, MANEUVERS
DAY		NIGHT		CROSS-COUNTRY		ACTUAL INSTR.		SIMULATED INSTR.						
														I certify that the statements made by me on this form are true.
														PILOT'S SIGNATURE

YEAR 20__ DATE	AIRCRAFT MAKE & MODEL	AIRCRAFT IDENT.	POINTS OF DEPARTURE & ARRIVAL		AIRCRAFT CATEGORY						GROUND TRAINER	TYPE OF PILOTING TIME			
			FROM	TO	AIRPLANE SEL		AIRPLANE MEL					DUAL RECEIVED	PILOT-IN-COMMAND		
				PAGE TOTAL											
				AMOUNT FORWARD											
				TOTAL TO DATE											

CONDITIONS OF FLIGHT										NO. INSTR. APPR.	NO. LDG. DAY / NIGHT		TOTAL DURATION OF FLIGHT		REMARKS, PROCEDURES, MANEUVERS
DAY		NIGHT		CROSS-COUNTRY		ACTUAL INSTR.		SIMULATED INSTR.							
															I certify that the statements made by me on this form are true.
															PILOT'S SIGNATURE

YEAR 20___ DATE	AIRCRAFT MAKE & MODEL	AIRCRAFT IDENT.	POINTS OF DEPARTURE & ARRIVAL		AIRCRAFT CATEGORY					GROUND TRAINER	TYPE OF PILOTING TIME				
			FROM	TO	AIRPLANE SEL		AIRPLANE MEL				DUAL RECEIVED		PILOT-IN-COMMAND		
			PAGE TOTAL												
			AMOUNT FORWARD												
			TOTAL TO DATE												

CONDITIONS OF FLIGHT											NO. INSTR. APPR.	NO. LDG. DAY / NIGHT	TOTAL DURATION OF FLIGHT		REMARKS, PROCEDURES, MANEUVERS
DAY		NIGHT		CROSS-COUNTRY		ACTUAL INSTR.		SIMULATED INSTR.							
															I certify that the statements made by me on this form are true.
															PILOT'S SIGNATURE

YEAR 20__ DATE	AIRCRAFT MAKE & MODEL	AIRCRAFT IDENT.	POINTS OF DEPARTURE & ARRIVAL		AIRCRAFT CATEGORY					GROUND TRAINER	TYPE OF PILOTING TIME			
			FROM	TO	AIRPLANE SEL		AIRPLANE MEL				DUAL RECEIVED		PILOT-IN-COMMAND	
PAGE TOTAL														
AMOUNT FORWARD														
TOTAL TO DATE														

CONDITIONS OF FLIGHT										NO. INSTR. APPR.	NO. LDG. DAY / NIGHT	TOTAL DURATION OF FLIGHT		REMARKS, PROCEDURES, MANEUVERS
DAY		NIGHT		CROSS-COUNTRY		ACTUAL INSTR.		SIMULATED INSTR.						
											/			
											/			
											/			
											/			
											/			
											/			
											/			
											/			*I certify that the statements made by me on this form are true.*
											/			
											/			PILOT'S SIGNATURE

YEAR 20___ DATE	AIRCRAFT MAKE & MODEL	AIRCRAFT IDENT.	POINTS OF DEPARTURE & ARRIVAL		AIRCRAFT CATEGORY			GROUND TRAINER	TYPE OF PILOTING TIME		
			FROM	TO	AIRPLANE SEL	AIRPLANE MEL			DUAL RECEIVED	PILOT-IN-COMMAND	
			PAGE TOTAL								
			AMOUNT FORWARD								
			TOTAL TO DATE								

CONDITIONS OF FLIGHT										NO. INSTR. APPR.	NO. LDG. DAY/NIGHT		TOTAL DURATION OF FLIGHT		REMARKS, PROCEDURES, MANEUVERS
DAY		NIGHT		CROSS-COUNTRY		ACTUAL INSTR.		SIMULATED INSTR.							
															I certify that the statements made by me on this form are true.
															PILOT'S SIGNATURE

YEAR 20___ DATE	AIRCRAFT MAKE & MODEL	AIRCRAFT IDENT.	POINTS OF DEPARTURE & ARRIVAL		AIRCRAFT CATEGORY				GROUND TRAINER		TYPE OF PILOTING TIME			
			FROM	TO	AIRPLANE SEL	AIRPLANE MEL					DUAL RECEIVED	PILOT-IN-COMMAND		
			PAGE TOTAL											
			AMOUNT FORWARD											
			TOTAL TO DATE											

CONDITIONS OF FLIGHT											NO. INSTR. APPR.	NO. LDG. DAY / NIGHT		TOTAL DURATION OF FLIGHT		REMARKS, PROCEDURES, MANEUVERS
DAY		NIGHT		CROSS-COUNTRY		ACTUAL INSTR.		SIMULATED INSTR.								
																I certify that the statements made by me on this form are true.
																PILOT'S SIGNATURE

YEAR 20___ DATE	AIRCRAFT MAKE & MODEL	AIRCRAFT IDENT.	POINTS OF DEPARTURE & ARRIVAL		AIRCRAFT CATEGORY				GROUND TRAINER	TYPE OF PILOTING TIME			
			FROM	TO	AIRPLANE SEL	AIRPLANE MEL				DUAL RECEIVED	PILOT-IN-COMMAND		
PAGE TOTAL													
AMOUNT FORWARD													
TOTAL TO DATE													

CONDITIONS OF FLIGHT											NO. INSTR. APPR.	NO. LDG. DAY / NIGHT	TOTAL DURATION OF FLIGHT		REMARKS, PROCEDURES, MANEUVERS
DAY		NIGHT		CROSS-COUNTRY		ACTUAL INSTR.		SIMULATED INSTR.							
															I certify that the statements made by me on this form are true.
															PILOT'S SIGNATURE

YEAR 20___ DATE	AIRCRAFT MAKE & MODEL	AIRCRAFT IDENT.	POINTS OF DEPARTURE & ARRIVAL		AIRCRAFT CATEGORY				GROUND TRAINER		TYPE OF PILOTING TIME					
			FROM	TO	AIRPLANE SEL		AIRPLANE MEL				DUAL RECEIVED		PILOT-IN-COMMAND			
			PAGE TOTAL													
			AMOUNT FORWARD													
			TOTAL TO DATE													

CONDITIONS OF FLIGHT										NO. INSTR. APPR.	NO. LDG. DAY / NIGHT		TOTAL DURATION OF FLIGHT		REMARKS, PROCEDURES, MANEUVERS
DAY		NIGHT		CROSS-COUNTRY		ACTUAL INSTR.		SIMULATED INSTR.							
															I certify that the statements made by me on this form are true.
															PILOT'S SIGNATURE

| YEAR 20___ DATE | AIRCRAFT MAKE & MODEL | AIRCRAFT IDENT. | POINTS OF DEPARTURE & ARRIVAL | | AIRCRAFT CATEGORY | | | GROUND TRAINER | TYPE OF PILOTING TIME | | | |
			FROM	TO	AIRPLANE SEL	AIRPLANE MEL			DUAL RECEIVED	PILOT-IN-COMMAND		
PAGE TOTAL												
AMOUNT FORWARD												
TOTAL TO DATE												

CONDITIONS OF FLIGHT											NO. INSTR. APPR.	NO. LDG. DAY/NIGHT		TOTAL DURATION OF FLIGHT		REMARKS, PROCEDURES, MANEUVERS
DAY		NIGHT		CROSS-COUNTRY		ACTUAL INSTR.		SIMULATED INSTR.								
																I certify that the statements made by me on this form are true.
																PILOT'S SIGNATURE

YEAR 20___ DATE	AIRCRAFT MAKE & MODEL	AIRCRAFT IDENT.	POINTS OF DEPARTURE & ARRIVAL		AIRCRAFT CATEGORY				GROUND TRAINER	TYPE OF PILOTING TIME			
			FROM	TO	AIRPLANE SEL	AIRPLANE MEL				DUAL RECEIVED	PILOT-IN-COMMAND		
PAGE TOTAL													
AMOUNT FORWARD													
TOTAL TO DATE													

CONDITIONS OF FLIGHT											NO. INSTR. APPR.	NO. LDG. DAY/NIGHT	TOTAL DURATION OF FLIGHT		REMARKS, PROCEDURES, MANEUVERS
DAY		NIGHT		CROSS-COUNTRY		ACTUAL INSTR.		SIMULATED INSTR.							
															I certify that the statements made by me on this form are true.
															PILOT'S SIGNATURE

YEAR 20__ DATE	AIRCRAFT MAKE & MODEL	AIRCRAFT IDENT.	POINTS OF DEPARTURE & ARRIVAL		AIRCRAFT CATEGORY					GROUND TRAINER	TYPE OF PILOTING TIME				
			FROM	TO	AIRPLANE SEL		AIRPLANE MEL				DUAL RECEIVED		PILOT-IN-COMMAND		
PAGE TOTAL															
AMOUNT FORWARD															
TOTAL TO DATE															

CONDITIONS OF FLIGHT											NO. INSTR. APPR.	NO. LDG. DAY/NIGHT		TOTAL DURATION OF FLIGHT		REMARKS, PROCEDURES, MANEUVERS
DAY		NIGHT		CROSS-COUNTRY		ACTUAL INSTR.		SIMULATED INSTR.								
																I certify that the statements made by me on this form are true.
																PILOT'S SIGNATURE

YEAR 20___ DATE	AIRCRAFT MAKE & MODEL	AIRCRAFT IDENT.	POINTS OF DEPARTURE & ARRIVAL		AIRCRAFT CATEGORY					GROUND TRAINER	TYPE OF PILOTING TIME					
			FROM	TO	AIRPLANE SEL		AIRPLANE MEL					DUAL RECEIVED		PILOT-IN-COMMAND		
			PAGE TOTAL													
			AMOUNT FORWARD													
			TOTAL TO DATE													

CONDITIONS OF FLIGHT										NO. INSTR. APPR.	NO. LDG. DAY / NIGHT		TOTAL DURATION OF FLIGHT		REMARKS, PROCEDURES, MANEUVERS
DAY		NIGHT		CROSS-COUNTRY		ACTUAL INSTR.		SIMULATED INSTR.							
															I certify that the statements made by me on this form are true.
															PILOT'S SIGNATURE

YEAR 20___ DATE	AIRCRAFT MAKE & MODEL	AIRCRAFT IDENT.	POINTS OF DEPARTURE & ARRIVAL		AIRCRAFT CATEGORY						GROUND TRAINER	TYPE OF PILOTING TIME					
			FROM	TO	AIRPLANE SEL		AIRPLANE MEL					DUAL RECEIVED		PILOT-IN-COMMAND			
				PAGE TOTAL													
				AMOUNT FORWARD													
				TOTAL TO DATE													

CONDITIONS OF FLIGHT											NO. INSTR. APPR.	NO. LDG.		TOTAL DURATION OF FLIGHT		REMARKS, PROCEDURES, MANEUVERS
DAY		NIGHT		CROSS-COUNTRY		ACTUAL INSTR.		SIMULATED INSTR.				DAY / NIGHT				
																I certify that the statements made by me on this form are true.
																PILOT'S SIGNATURE

YEAR 20___ DATE	AIRCRAFT MAKE & MODEL	AIRCRAFT IDENT.	POINTS OF DEPARTURE & ARRIVAL		AIRCRAFT CATEGORY						GROUND TRAINER		TYPE OF PILOTING TIME					
			FROM	TO	AIRPLANE SEL		AIRPLANE MEL						DUAL RECEIVED		PILOT-IN-COMMAND			
				PAGE TOTAL														
				AMOUNT FORWARD														
				TOTAL TO DATE														

CONDITIONS OF FLIGHT											NO. INSTR. APPR.	NO. LDG. DAY / NIGHT		TOTAL DURATION OF FLIGHT		REMARKS, PROCEDURES, MANEUVERS
DAY		NIGHT		CROSS-COUNTRY		ACTUAL INSTR.		SIMULATED INSTR.								
																I certify that the statements made by me on this form are true.
																PILOT'S SIGNATURE

YEAR 20___ DATE	AIRCRAFT MAKE & MODEL	AIRCRAFT IDENT.	POINTS OF DEPARTURE & ARRIVAL		AIRCRAFT CATEGORY				GROUND TRAINER	TYPE OF PILOTING TIME			
			FROM	TO	AIRPLANE SEL	AIRPLANE MEL				DUAL RECEIVED	PILOT-IN-COMMAND		
PAGE TOTAL													
AMOUNT FORWARD													
TOTAL TO DATE													

CONDITIONS OF FLIGHT								NO. INSTR. APPR.	NO. LDG. DAY/NIGHT	TOTAL DURATION OF FLIGHT	REMARKS, PROCEDURES, MANEUVERS
DAY		NIGHT		CROSS-COUNTRY		ACTUAL INSTR.		SIMULATED INSTR.			
														I certify that the statements made by me on this form are true.
														PILOT'S SIGNATURE

YEAR 20___ DATE	AIRCRAFT MAKE & MODEL	AIRCRAFT IDENT.	POINTS OF DEPARTURE & ARRIVAL		AIRCRAFT CATEGORY			GROUND TRAINER	TYPE OF PILOTING TIME		
			FROM	TO	AIRPLANE SEL	AIRPLANE MEL			DUAL RECEIVED	PILOT-IN-COMMAND	
PAGE TOTAL											
AMOUNT FORWARD											
TOTAL TO DATE											

CONDITIONS OF FLIGHT											NO. INSTR. APPR.	NO. LDG. DAY / NIGHT	TOTAL DURATION OF FLIGHT		REMARKS, PROCEDURES, MANEUVERS
DAY		NIGHT		CROSS-COUNTRY		ACTUAL INSTR.		SIMULATED INSTR.							
															I certify that the statements made by me on this form are true.
															PILOT'S SIGNATURE

| YEAR 20__ DATE | AIRCRAFT MAKE & MODEL | AIRCRAFT IDENT. | POINTS OF DEPARTURE & ARRIVAL | | AIRCRAFT CATEGORY | | | | | GROUND TRAINER | | TYPE OF PILOTING TIME | | | |
			FROM	TO	AIRPLANE SEL		AIRPLANE MEL					DUAL RECEIVED		PILOT-IN-COMMAND	
PAGE TOTAL															
AMOUNT FORWARD															
TOTAL TO DATE															

CONDITIONS OF FLIGHT											NO. INSTR. APPR.	NO. LDG. DAY / NIGHT	TOTAL DURATION OF FLIGHT		REMARKS, PROCEDURES, MANEUVERS
DAY		NIGHT		CROSS-COUNTRY		ACTUAL INSTR.		SIMULATED INSTR.							
															I certify that the statements made by me on this form are true.
															PILOT'S SIGNATURE

YEAR 20___ DATE	AIRCRAFT MAKE & MODEL	AIRCRAFT IDENT.	POINTS OF DEPARTURE & ARRIVAL		AIRCRAFT CATEGORY				GROUND TRAINER		TYPE OF PILOTING TIME			
			FROM	TO	AIRPLANE SEL	AIRPLANE MEL					DUAL RECEIVED	PILOT-IN-COMMAND		
PAGE TOTAL														
AMOUNT FORWARD														
TOTAL TO DATE														

CONDITIONS OF FLIGHT										NO. INSTR. APPR.	NO. LDG. DAY / NIGHT	TOTAL DURATION OF FLIGHT		REMARKS, PROCEDURES, MANEUVERS
DAY		NIGHT		CROSS-COUNTRY		ACTUAL INSTR.		SIMULATED INSTR.						
														I certify that the statements made by me on this form are true.
														PILOT'S SIGNATURE

YEAR 20___ DATE	AIRCRAFT MAKE & MODEL	AIRCRAFT IDENT.	POINTS OF DEPARTURE & ARRIVAL		AIRCRAFT CATEGORY					GROUND TRAINER		TYPE OF PILOTING TIME					
			FROM	TO	AIRPLANE SEL		AIRPLANE MEL					DUAL RECEIVED		PILOT-IN-COMMAND			
PAGE TOTAL																	
AMOUNT FORWARD																	
TOTAL TO DATE																	

CONDITIONS OF FLIGHT												NO. INSTR. APPR.	NO. LDG. DAY / NIGHT		TOTAL DURATION OF FLIGHT		REMARKS, PROCEDURES, MANEUVERS
DAY		NIGHT		CROSS-COUNTRY		ACTUAL INSTR.		SIMULATED INSTR.									
																	I certify that the statements made by me on this form are true.
																	PILOT'S SIGNATURE

YEAR 20___ DATE	AIRCRAFT MAKE & MODEL	AIRCRAFT IDENT.	POINTS OF DEPARTURE & ARRIVAL		AIRCRAFT CATEGORY					GROUND TRAINER	TYPE OF PILOTING TIME			
			FROM	TO	AIRPLANE SEL		AIRPLANE MEL				DUAL RECEIVED		PILOT-IN-COMMAND	
PAGE TOTAL														
AMOUNT FORWARD														
TOTAL TO DATE														

CONDITIONS OF FLIGHT										NO. INSTR. APPR.	NO. LDG. DAY / NIGHT	TOTAL DURATION OF FLIGHT		REMARKS, PROCEDURES, MANEUVERS
DAY		NIGHT		CROSS-COUNTRY		ACTUAL INSTR.		SIMULATED INSTR.						
														I certify that the statements made by me on this form are true.
														PILOT'S SIGNATURE

| YEAR 20___ DATE | AIRCRAFT MAKE & MODEL | AIRCRAFT IDENT. | POINTS OF DEPARTURE & ARRIVAL | | AIRCRAFT CATEGORY | | | | | GROUND TRAINER | TYPE OF PILOTING TIME | | | |
			FROM	TO	AIRPLANE SEL		AIRPLANE MEL				DUAL RECEIVED	PILOT-IN-COMMAND		
				PAGE TOTAL										
				AMOUNT FORWARD										
				TOTAL TO DATE										

CONDITIONS OF FLIGHT										NO. INSTR. APPR.	NO. LDG. DAY / NIGHT	TOTAL DURATION OF FLIGHT		REMARKS, PROCEDURES, MANEUVERS
DAY		NIGHT		CROSS-COUNTRY		ACTUAL INSTR.		SIMULATED INSTR.						
														I certify that the statements made by me on this form are true.
														PILOT'S SIGNATURE

YEAR 20___ DATE	AIRCRAFT MAKE & MODEL	AIRCRAFT IDENT.	POINTS OF DEPARTURE & ARRIVAL		AIRCRAFT CATEGORY				GROUND TRAINER	TYPE OF PILOTING TIME			
			FROM	TO	AIRPLANE SEL	AIRPLANE MEL				DUAL RECEIVED	PILOT-IN-COMMAND		
PAGE TOTAL													
AMOUNT FORWARD													
TOTAL TO DATE													

CONDITIONS OF FLIGHT										NO. INSTR. APPR.	NO. LDG. DAY / NIGHT	TOTAL DURATION OF FLIGHT		REMARKS, PROCEDURES, MANEUVERS
DAY		NIGHT		CROSS-COUNTRY		ACTUAL INSTR.		SIMULATED INSTR.						
														I certify that the statements made by me on this form are true.
														PILOT'S SIGNATURE

YEAR 20___ DATE	AIRCRAFT MAKE & MODEL	AIRCRAFT IDENT.	POINTS OF DEPARTURE & ARRIVAL		AIRCRAFT CATEGORY			GROUND TRAINER	TYPE OF PILOTING TIME			
			FROM	TO	AIRPLANE SEL	AIRPLANE MEL			DUAL RECEIVED	PILOT-IN-COMMAND		
				PAGE TOTAL								
				AMOUNT FORWARD								
				TOTAL TO DATE								

CONDITIONS OF FLIGHT										NO. INSTR. APPR.	NO. LDG. DAY / NIGHT	TOTAL DURATION OF FLIGHT		REMARKS, PROCEDURES, MANEUVERS
DAY		NIGHT		CROSS-COUNTRY		ACTUAL INSTR.		SIMULATED INSTR.						
											/			
											/			
											/			
											/			
											/			
											/			
											/			
											/			*I certify that the statements made by me on this form are true.*
											/			
											/			PILOT'S SIGNATURE

YEAR 20___ DATE	AIRCRAFT MAKE & MODEL	AIRCRAFT IDENT.	POINTS OF DEPARTURE & ARRIVAL		AIRCRAFT CATEGORY				GROUND TRAINER	TYPE OF PILOTING TIME			
			FROM	TO	AIRPLANE SEL		AIRPLANE MEL			DUAL RECEIVED		PILOT-IN-COMMAND	
			PAGE TOTAL										
			AMOUNT FORWARD										
			TOTAL TO DATE										

CONDITIONS OF FLIGHT											NO. INSTR. APPR.	NO. LDG. DAY / NIGHT		TOTAL DURATION OF FLIGHT		REMARKS, PROCEDURES, MANEUVERS
DAY		NIGHT		CROSS-COUNTRY		ACTUAL INSTR.		SIMULATED INSTR.								
																I certify that the statements made by me on this form are true.
																PILOT'S SIGNATURE

| YEAR 20__ DATE | AIRCRAFT MAKE & MODEL | AIRCRAFT IDENT. | POINTS OF DEPARTURE & ARRIVAL | | AIRCRAFT CATEGORY | | | | GROUND TRAINER | TYPE OF PILOTING TIME | | | |
			FROM	TO	AIRPLANE SEL	AIRPLANE MEL				DUAL RECEIVED	PILOT-IN-COMMAND		
PAGE TOTAL													
AMOUNT FORWARD													
TOTAL TO DATE													

CONDITIONS OF FLIGHT											NO. INSTR. APPR.	NO. LDG. DAY/NIGHT		TOTAL DURATION OF FLIGHT		REMARKS, PROCEDURES, MANEUVERS
DAY		NIGHT		CROSS-COUNTRY		ACTUAL INSTR.		SIMULATED INSTR.				DAY	NIGHT			
																I certify that the statements made by me on this form are true.
																PILOT'S SIGNATURE

YEAR 20___ DATE	AIRCRAFT MAKE & MODEL	AIRCRAFT IDENT.	POINTS OF DEPARTURE & ARRIVAL		AIRCRAFT CATEGORY					GROUND TRAINER	TYPE OF PILOTING TIME					
			FROM	TO	AIRPLANE SEL		AIRPLANE MEL					DUAL RECEIVED		PILOT-IN-COMMAND		
PAGE TOTAL																
AMOUNT FORWARD																
TOTAL TO DATE																

CONDITIONS OF FLIGHT										NO. INSTR. APPR.	NO. LDG. DAY/NIGHT	TOTAL DURATION OF FLIGHT		REMARKS, PROCEDURES, MANEUVERS
DAY		NIGHT		CROSS-COUNTRY		ACTUAL INSTR.		SIMULATED INSTR.						
														I certify that the statements made by me on this form are true.
														PILOT'S SIGNATURE

YEAR 20___ DATE	AIRCRAFT MAKE & MODEL	AIRCRAFT IDENT.	POINTS OF DEPARTURE & ARRIVAL		AIRCRAFT CATEGORY				GROUND TRAINER	TYPE OF PILOTING TIME			
			FROM	TO	AIRPLANE SEL	AIRPLANE MEL				DUAL RECEIVED	PILOT-IN-COMMAND		
PAGE TOTAL													
AMOUNT FORWARD													
TOTAL TO DATE													

CONDITIONS OF FLIGHT										NO. INSTR. APPR.	NO. LDG. DAY / NIGHT		TOTAL DURATION OF FLIGHT		REMARKS, PROCEDURES, MANEUVERS
DAY		NIGHT		CROSS-COUNTRY		ACTUAL INSTR.		SIMULATED INSTR.							
															I certify that the statements made by me on this form are true.
															PILOT'S SIGNATURE

YEAR 20___ DATE	AIRCRAFT MAKE & MODEL	AIRCRAFT IDENT.	POINTS OF DEPARTURE & ARRIVAL		AIRCRAFT CATEGORY						GROUND TRAINER		TYPE OF PILOTING TIME					
			FROM	TO	AIRPLANE SEL		AIRPLANE MEL						DUAL RECEIVED		PILOT-IN-COMMAND			
PAGE TOTAL																		
AMOUNT FORWARD																		
TOTAL TO DATE																		

CONDITIONS OF FLIGHT											NO. INSTR. APPR.	NO. LDG. DAY / NIGHT	TOTAL DURATION OF FLIGHT		REMARKS, PROCEDURES, MANEUVERS
DAY		NIGHT		CROSS-COUNTRY		ACTUAL INSTR.		SIMULATED INSTR.							
															I certify that the statements made by me on this form are true.
															PILOT'S SIGNATURE

YEAR 20___ DATE	AIRCRAFT MAKE & MODEL	AIRCRAFT IDENT.	POINTS OF DEPARTURE & ARRIVAL		AIRCRAFT CATEGORY					GROUND TRAINER	TYPE OF PILOTING TIME					
			FROM	TO	AIRPLANE SEL		AIRPLANE MEL				DUAL RECEIVED		PILOT-IN-COMMAND			
PAGE TOTAL																
AMOUNT FORWARD																
TOTAL TO DATE																

CONDITIONS OF FLIGHT										NO. INSTR. APPR.	NO. LDG.		TOTAL DURATION OF FLIGHT		REMARKS, PROCEDURES, MANEUVERS
DAY		NIGHT		CROSS-COUNTRY		ACTUAL INSTR.		SIMULATED INSTR.			DAY	NIGHT			
															I certify that the statements made by me on this form are true.
															PILOT'S SIGNATURE

YEAR 20___ DATE	AIRCRAFT MAKE & MODEL	AIRCRAFT IDENT.	POINTS OF DEPARTURE & ARRIVAL		AIRCRAFT CATEGORY					GROUND TRAINER		TYPE OF PILOTING TIME					
			FROM	TO	AIRPLANE SEL		AIRPLANE MEL					DUAL RECEIVED		PILOT-IN-COMMAND			
			PAGE TOTAL														
			AMOUNT FORWARD														
			TOTAL TO DATE														

CONDITIONS OF FLIGHT										NO. INSTR. APPR.	NO. LDG. DAY /NIGHT	TOTAL DURATION OF FLIGHT		REMARKS, PROCEDURES, MANEUVERS
DAY		NIGHT		CROSS-COUNTRY		ACTUAL INSTR.		SIMULATED INSTR.						
														I certify that the statements made by me on this form are true.
														PILOT'S SIGNATURE

GROUND INSTRUCTION LOG				
DATE	LESSON PLAN	INSTRUCTOR	TIME	TOTAL

GROUND INSTRUCTION LOG				
DATE	LESSON PLAN	INSTRUCTOR	TIME	TOTAL

YEAR 20___	CLASSIFICATION OF PILOT-IN-COMMAND TIME												
	JAN.	FEB.	MAR.	APR.	MAY	JUNE	JULY	AUG.	SEPT.	OCT.	NOV.	DEC.	TOTAL
SINGLE ENGINE													
MULTI-ENGINE													
TURBINE ENGINE													
CROSS-COUNTRY													
NIGHT													
GROUND TRAINER													
SIMULATED INSTRUMENT													
ACTUAL INSTRUMENT													
UNMANNED (UAS)													

YEAR 20___	CLASSIFICATION OF PILOT-IN-COMMAND TIME												
	JAN.	FEB.	MAR.	APR.	MAY	JUNE	JULY	AUG.	SEPT.	OCT.	NOV.	DEC.	TOTAL
SINGLE ENGINE													
MULTI-ENGINE													
TURBINE ENGINE													
CROSS-COUNTRY													
NIGHT													
GROUND TRAINER													
SIMULATED INSTRUMENT													
ACTUAL INSTRUMENT													
UNMANNED (UAS)													

AIRCRAFT FLOWN AND NUMBER OF HOURS IN EACH					
AIRCRAFT MAKE AND MODEL	PIC	DUAL REC'D.	AIRCRAFT MAKE AND MODEL	PIC	DUAL REC'D.

INITIAL SOLO ENDORSEMENTS

I certify that *(First name, MI, Last name)* _____,
has satisfactorily completed the presolo knowledge exam of §61.87(b), received the required presolo training required by §61.87(c), and has demonstrated the proficiency of §61.87(d) and is proficient to make solo flights in *(category, make and model aircraft)* _____.

SIGNED _____ DATE _____

CFI NO. _____ EXPIRATION DATE _____

I certify that *(First name, MI, Last name)* _____,
has received the required training to qualify for local solo flying. I have determined he/she meets the applicable requirements of §61.87(n) and is proficient to make solo flights in *(category, make and model aircraft)* _____ until *(maximum 90 days from date given)* _____.
Limitations: _____

SIGNED _____ DATE _____

CFI NO. _____ EXPIRATION DATE _____

ADDITIONAL AIRPORT WITHIN 25NM SOLO ENDORSEMENTS

I certify that *(First name, MI, Last name)* _____,
has received the required training of § 61.93(b)(1). I have determined that he/she is proficient to practice solo takeoffs and landings at *(airport name)* _____ subject to the following conditions:
_____.

SIGNED _____ DATE _____

CFI NO. _____ EXPIRATION DATE _____

I certify that *(First name, MI, Last name)* _____,
has received the required training of § 61.93(b)(1). I have determined that he/she is proficient to practice solo takeoffs and landings at *(airport name)* _____ subject to the following conditions:
_____.

SIGNED _____ DATE _____

CFI NO. _____ EXPIRATION DATE _____

CROSS-COUNTRY SOLO ENDORSEMENTS

I certify that *(First name, MI, Last name)* _____,
has received the required solo cross-country training and find he/she has met the applicable requirements of §61.93, and is proficient to make solo cross-country flights in a *(category, make and model aircraft)* _____.

SIGNED _____ DATE _____

CFI NO. _____ EXPIRATION DATE _____

I have reviewed the cross-country planning of *(First name, MI, Last name)* _____ _____ and find the planning and preparation to be correct to make the solo flight from *(location)* _____ to *(destination)* _____ via *(route of flight)* _____ with landings at *(name the airports)* _____ in a *(make and model aircraft)* _____ on *(date)* _____.
List any applicable conditions or limitations: _____

SIGNED _____ DATE _____

CFI NO. _____ EXPIRATION DATE _____

ADDITIONAL SOLO ENDORSEMENTS

90-DAY SOLO
I certify that *(First name, MI, Last name)* _____,
has received the required training to qualify for local solo flying. I have determined he/she meets the applicable requirements of §61.87(n) and is proficient to make solo flights in *(category, make and model aircraft)* _____ _____ until *(max 90 days from date given)* _____.
Limitations: _____.
SIGNED _____ DATE _____
CFI NO. _____ EXPIRATION DATE _____

I certify that *(First name, MI, Last name)* _____,
has received the required training to qualify for local solo flying. I have determined he/she meets the applicable requirements of §61.87(n) and is proficient to make solo flights in *(category, make and model aircraft)* _____ _____ until *(max 90 days from date given)* _____.
Limitations: _____.
SIGNED _____ DATE _____
CFI NO. _____ EXPIRATION DATE _____

NIGHT SOLO
I certify that *(First name, MI, Last name)* _____,
has received the required presolo training in a *(category, make and model aircraft)* _____ and determined he/she has demonstrated the proficiency of §61.87(o) and is proficient to make solo flights at night in a *(category, make and model aircraft)* _____,
SIGNED _____ DATE _____
CFI NO. _____ EXPIRATION DATE _____

CLASS B SOLO
I certify that *(First name, MI, Last name)* _____,
has received the required training of §61.95(a) and determined he/she is proficient to conduct solo flights in *(name of Class B)* _____ airspace. *List any applicable conditions or limitations:* _____
_____.
SIGNED _____ DATE _____
CFI NO. _____ EXPIRATION DATE _____

I certify that *(First name, MI, Last name)* _____,
has received the required training of §61.95(a)(1) and determined that he/she is proficient to conduct solo flight operations at *(name of airport)* _____
_____. *List any applicable conditions or limitations:* _____
_____.
SIGNED _____ DATE _____
CFI NO. _____ EXPIRATION DATE _____

REPEAT X/C SOLO (LESS THAN 50 NM)
I certify that *(First name, MI, Last name)* _____,
has received the required training in both directions between and at both *(airport names)* _____ and determined that he/she is proficient in §61.93(b)(2) to conduct repeated solo cross-country flights over that route, which is not more than 50NM from the point of departure, subject to the following conditions: _____.
SIGNED _____ DATE _____
CFI NO. _____ EXPIRATION DATE _____

INITIAL SPORT PILOT ENDORSEMENTS

SOLO FLIGHT 1st 90-DAY

I certify that *(First name, MI, Last name)* _____,
has received the required training to qualify for local solo flying. I have
determined he/she meets the applicable requirements of §61.87(n) and is
proficient to make solo flights in *(make and model aircraft)* _____
until *(maximum 90 days from date given)* _____.
Limitations: _____

SIGNED _____ DATE _____

CFI NO. _____ EXPIRATION DATE _____

SOLO FLIGHT IN CLASS B, C, AND D AIRSPACE *(req. each add'l 90-day period)*

I certify that *(First name, MI, Last name)* _____,
has received the required training of §61.94(a). I have determined he/she is
proficient to conduct solo flights in *(name of Class B, C, or D)* ____ airspace.
List any applicable conditions or limitations: _____

SIGNED _____ DATE _____

CFI NO. _____ EXPIRATION DATE _____

TAKING THE AERONAUTICAL KNOWLEDGE TEST

I certify that *(First name, MI, Last name)* _____,
has received the required aeronautical knowledge training of §61.309. I have
determined that he/she is prepared for the *(name the knowledge test/aircraft
category)* _____.

SIGNED _____ DATE _____

CFI NO. _____ EXPIRATION DATE _____

TAKING THE SPORT PILOT PRACTICAL TEST

I certify that *(First name, MI, Last name)* _____,
has received the required training of §§61.309 and 61.311 and met the expe-
rience requirements of §61.313. I have determined that he/she is prepared
for the *(category/class)* _____ practical test.

SIGNED _____ DATE _____

CFI NO. _____ EXPIRATION DATE _____

I certify that *(First name, MI, Last name)* _____,
has received the required training and demonstrated satisfactory knowledge
of deficient areas from the Sport Pilot knowledge test as required per §61.39
for the practical test.

SIGNED _____ DATE _____

CFI NO. _____ EXPIRATION DATE _____

PASSING THE SPORT PILOT PRACTICAL TEST

I certify that *(First name, MI, Last name)* _____,
has met the training and endorsement requirements of §§61.309, 61.311 and
61.313. I have determined him/her proficient to act as PIC of *(category/class)*
_____ of light-sport aircraft.

SIGNED _____ DATE _____

PILOT CERT # _____ DPE # _____ EXP. DATE _____

ADDITIONAL SPORT PILOT ENDORSEMENTS

TAKING FLIGHT PROFICIENCY CHECK FOR AN ADDITIONAL AIRCRAFT

I certify that *(First name, MI, Last name)* _____,
has received the required training of §§61.309 and 61.311. I have determined
that he/she is prepared for the *(category / class)* _____
proficiency check.

SIGNED _____ DATE _____

CFI NO. _____ EXPIRATION DATE _____

PASSING FLIGHT PROFICIENCY CHECK FOR AN ADDITIONAL AIRCRAFT

I certify that *(First name, MI, Last name)* _____,
has met the training and endorsement requirements of §§61.309 and 61.311.
I have determined him/her proficient to act as PIC of *(category / class)* _____
_____ of light-sport aircraft.

SIGNED _____ DATE _____

CFI NO. _____ EXPIRATION DATE _____

PRIVILEGES TO OPERATE A DIFFERENT AIRCRAFT

I certify that *(First name, MI, Last name)* _____,
has received the required training of §61.321 in a *(category / class)* _____
_____. I have determined him/her proficient to act as PIC
of that light-sport aircraft.

SIGNED _____ DATE _____

CFI NO. _____ EXPIRATION DATE _____

AIRSPACE & AIRSPEED ENDORSEMENTS

I certify that *(First name, MI, Last name)* _____,
has received the required training of §61.327 in a *(make and model aircraft)*
_____. I have determined him/her proficient to act
as PIC of a light-sport aircraft that has a V_H greater/less than 87 knots (kts) CAS.

SIGNED _____ DATE _____

CFI NO. _____ EXPIRATION DATE _____

I certify that *(First name, MI, Last name)* _____,
has received the required training of §61.325. I have determined he/she is
proficient to conduct operations in Class B, C, or D airspace, at an airport
located in Class B, C, or D airspace, or to, from, through, or on an airport
having an operational control tower.

SIGNED _____ DATE _____

CFI NO. _____ EXPIRATION DATE _____

I certify that *(First name, MI, Last name)* _____,
has received the training required in accordance with §61.327 in a *(make and
model aircraft)* . I have determined him/her proficient to act as PIC of a light-
sport aircraft that has a V_H greater/less than 87 knots (kts) CAS.

SIGNED _____ DATE _____

CFI NO. _____ EXPIRATION DATE _____

RECREATIONAL PILOT ENDORSEMENTS

I certify that *(First name, MI, Last name)* _____,
has received the required training of §61.97(b) and/or I have reviewed the
home study curriculum and have determined that he/she is prepared for the
(name the knowledge test / aircraft category) _____.

SIGNED _____ DATE _____

CFI NO. _____ EXPIRATION DATE _____

I certify that *(First name, MI, Last name)* _____,
has received the required training of §§61.98(b) and 61.99 and have determined
that he/she is prepared for the *(name the practical test)* _____

_____.

SIGNED _____ DATE _____

CFI NO. _____ EXPIRATION DATE _____

I certify that *(First name, MI, Last name)* _____,
has received the required training of §61.101(b) and have determined he/she
is competent to operate at the *(name of airport)* _____.

SIGNED _____ DATE _____

CFI NO. _____ EXPIRATION DATE _____

I certify that *(First name, MI, Last name)* _____,
has received the required cross-country training of §61.101(c) and have
determined that he/she is proficient in cross-country flying of part 61, subpart E.

SIGNED _____ DATE _____

CFI NO. _____ EXPIRATION DATE _____

I certify that *(First name, MI, Last name)* _____,
has received the required 180-day recurrent training of §61.101(f) in a *(make
and model aircraft)* _____.
I have determined him/her to be proficient to act as PIC of that aircraft.

SIGNED _____ DATE _____

CFI NO. _____ EXPIRATION DATE _____

I certify that *(First name, MI, Last name)* _____,
has received the required training of §61.87 in a *(make and model aircraft)*
_____. I have determined he/she is
prepared to conduct a solo flight on *(date)* under the following conditions: __

_____.

SIGNED _____ DATE _____

CFI NO. _____ EXPIRATION DATE _____

PRIVATE PILOT ENDORSEMENTS	COMMERCIAL PILOT ENDORSEMENTS
I certify that *(First name, MI, Last name)* _____, has received the required training of §61.105 and/or I have reviewed the home study curriculum and have determined he/she is prepared for the *(name the knowledge test / aircraft category)* _____. SIGNED _____ DATE _____ CFI NO. _____ EXPIRATION DATE _____	I certify that *(First name, MI, Last name)* _____, has received the required training of §61.125 and/or I have reviewed the home study curriculum and have determined that he/she is prepared for the *(name the knowledge test / aircraft category)* _____. SIGNED _____ DATE _____ CFI NO. _____ EXPIRATION DATE _____
I certify that *(First name, MI, Last name)* _____, has received the required training of §§61.107 and 61.109 and have determined he/she is prepared for the *(name the practical test)* _____. SIGNED _____ DATE _____ CFI NO. _____ EXPIRATION DATE _____	I certify that *(First name, MI, Last name)* _____, has received the required training of §§61.127 and 61.129 and have determined he/she is prepared for the *(name the practical test)* _____. SIGNED _____ DATE _____ CFI NO. _____ EXPIRATION DATE _____
I certify that *(First name, MI, Last name)* _____, has received the training as required by §61.39(a)(6)(i) within the preceding two calendar months and have determined that he/she is prepared for the *(name the practical test)* _____ and has demonstrated satisfactory knowledge of subject areas shown to be deficient on his/her Airman Knowledge Test as required by §61.39(a)(6)(iii). SIGNED _____ DATE _____ CFI NO. _____ EXPIRATION DATE _____	I certify that *(First name, MI, Last name)* _____, has received the training as required by §61.39(a)(6)(i) within the preceding two calendar months and have determined that he/she is prepared for the *(name the practical test)* _____ and has demonstrated satisfactory knowledge of subject areas shown to be deficient on his/her Airman Knowledge Test as required by §61.39(a)(6)(iii). SIGNED _____ DATE _____ CFI NO. _____ EXPIRATION DATE _____

INSTRUMENT RATING ENDORSEMENTS	INSTRUMENT PROFICIENCY
I certify that *(First name, MI, Last name)* _____, has received the required training of §61.65(b) and/or I have reviewed the home study curriculum and have determined that he/she is prepared for the *(name the knowledge test / aircraft category)* _____. SIGNED _____ DATE _____ CFI NO. _____ EXPIRATION DATE _____	I certify that *(First name, MI, Last name)* _____, *(pilot certificate)* _____, *(certificate number)* _____, has satisfactorily completed the instrument proficiency check required in §61.57(d) in a *(make and model of aircraft)* _____ on *(date)* _____. SIGNED _____ DATE _____ CFI NO. _____ EXPIRATION DATE _____
I certify that *(First name, MI, Last name)* _____, has received the required training of §61.65(c) and (d) and have determined he/she is prepared for the *(name of the practical test)* _____. SIGNED _____ DATE _____ CFI NO. _____ EXPIRATION DATE _____	I certify that *(First name, MI, Last name)* _____, *(pilot certificate)* _____, *(certificate number)* _____, has satisfactorily completed the instrument proficiency check required in §61.57(d) in a *(make and model of aircraft)* _____ on *(date)* _____. SIGNED _____ DATE _____ CFI NO. _____ EXPIRATION DATE _____
I certify that *(First name, MI, Last name)* _____, has received the training as required by §61.39(a)(6)(i) within the preceding two calendar months and have determined that he/she is prepared for the *(name the practical test)* _____ and has demonstrated satisfactory knowledge of subject areas shown to be deficient on his/her Airman Knowledge Test as required by §61.39(a)(6)(iii). SIGNED _____ DATE _____ CFI NO. _____ EXPIRATION DATE _____	I certify that *(First name, MI, Last name)* _____, *(pilot certificate)* _____, *(certificate number)* _____, has satisfactorily completed the instrument proficiency check required in §61.57(d) in a *(make and model of aircraft)* _____ on *(date)* _____. SIGNED _____ DATE _____ CFI NO. _____ EXPIRATION DATE _____

FLIGHT AND GROUND INSTRUCTOR ENDORSEMENTS

I certify that *(First name, MI, Last name)* _____,
has received the required fundamentals of instruction training of §61.185(a)(1).

SIGNED _____ DATE _____

CFI NO. _____ EXPIRATION DATE _____

I certify that *(First name, MI, Last name)* _____
has received the required training of §61.183(i) and have determined that he/she
is competent and proficient in instructional skills for training stall awareness,
spin entry, spins, and spin recovery procedures. *(Required of flight instructor
applicants for the airplane and glider ratings only.)*

SIGNED _____ DATE _____

CFI NO. _____ EXPIRATION DATE _____

I certify that *(First name, MI, Last name)* _____,
has received the required training of §61.187(b) and/or I have reviewed the
home study curriculum and have determined he/she is prepared for the *(name
the knowledge test / aircraft category)* _____.

SIGNED _____ DATE _____

CFI NO. _____ EXPIRATION DATE _____

I certify that *(First name, MI, Last name)* _____
has demonstrated satisfactory proficiency on the appropriate ground instructor
knowledge and training subjects of §61.213(a)(3) and (a)(4).

SIGNED _____ DATE _____

CFI NO. _____ EXPIRATION DATE _____

I certify that *(First name, MI, Last name)* _____,
has received the required CFII training of §61.187(b)(7) and have determined
he/she is prepared for the *(name of practical test)* _____

_____.

SIGNED _____ DATE _____

CFI NO. _____ EXPIRATION DATE _____

I certify that *(First name, MI, Last name)* _____,
has received the training as required by §61.39(a)(6)(i) within the preceding two
calendar months and have determined that he/she is prepared for the *(name
the practical test)* _____
and has demonstrated satisfactory knowledge of subject areas shown to be
deficient on his/her Airman Knowledge Test as required by §61.39(a)(6)(iii).

SIGNED _____ DATE _____

CFI NO. _____ EXPIRATION DATE _____

ADDITIONAL ENDORSEMENTS

COMPLEX AIRPLANE

I certify that *(First name, MI, Last name)* _____,
(pilot certificate) _____, *(certificate number)* _____,
has received the required training of §61.31(e) in a *(make and model of complex airplane)* _____. I have determined that he/she is proficient in the operation and systems of a complex airplane.

SIGNED _____ DATE _____

CFI NO. _____ EXPIRATION DATE _____

HIGH-PERFORMANCE AIRPLANE

I certify that *(First name, MI, Last name)* _____,
(pilot certificate) _____, *(certificate number)* _____,
has received the required training of §61.31(f) in a *(make and model of high performance airplane)* _____. I have determined that he/she is proficient in the operation and systems of a high performance airplane.

SIGNED _____ DATE _____

CFI NO. _____ EXPIRATION DATE _____

HIGH ALTITUDE

I certify that *(First name, MI, Last name)* _____,
(pilot certificate) _____, *(certificate number)* _____,
has received the required training of §61.31(g) in a *(make and model of pressurized aircraft)* _____. I have determined that he/she is proficient in the operation and systems of a pressurized aircraft.

SIGNED _____ DATE _____

CFI NO. _____ EXPIRATION DATE _____

TAILWHEEL

I certify that *(First name, MI, Last name)* _____,
(pilot certificate) _____, *(certificate number)* _____,
has received the required training of §61.31(i) in a *(make and model of tailwheel airplane)* _____. I have determined that he/she is proficient in the operation of a tailwheel airplane.

SIGNED _____ DATE _____

CFI NO. _____ EXPIRATION DATE _____

TYPE RATING

I certify that *(First name, MI, Last name)* _____,
has received the training as required by §61.31(d)(3) to serve as a PIC in a *(category and class of aircraft)* _____. I have determined that he/she is prepared to serve as PIC in that *(make and model of aircraft)* _____.

SIGNED _____ DATE _____

CFI NO. _____ EXPIRATION DATE _____

ADDED RATING

I certify that *(First name, MI, Last name)* _____,
(pilot certificate) _____, *(certificate number)* _____,
has received the required training for an additional *(name the aircraft category/class rating)* _____. I have determined that he/she is prepared for the *(name the practical test)* _____ for the addition of a *(name the aircraft category/class rating)* _____.

SIGNED _____ DATE _____

CFI NO. _____ EXPIRATION DATE _____

FLIGHT REVIEW

I certify that *(First name, MI, Last name)* _____,
(pilot certificate) _____, *(certificate number)* _____,
has satisfactorily completed the flight review required in §61.56(a) on *(date)*

_____.

SIGNED _____ DATE _____
CFI NO. _____ EXPIRATION DATE _____

I certify that *(First name, MI, Last name)* _____,
(pilot certificate) _____, *(certificate number)* _____,
has satisfactorily completed the flight review required in §61.56(a) on *(date)*

_____.

SIGNED _____ DATE _____
CFI NO. _____ EXPIRATION DATE _____

I certify that *(First name, MI, Last name)* _____,
(pilot certificate) _____, *(certificate number)* _____,
has satisfactorily completed the flight review required in §61.56(a) on *(date)*

_____.

SIGNED _____ DATE _____
CFI NO. _____ EXPIRATION DATE _____

I certify that *(First name, MI, Last name)* _____,
(pilot certificate) _____, *(certificate number)* _____,
has satisfactorily completed the flight review required in §61.56(a) on *(date)*

_____.

SIGNED _____ DATE _____
CFI NO. _____ EXPIRATION DATE _____

I certify that *(First name, MI, Last name)* _____,
(pilot certificate) _____, *(certificate number)* _____,
has satisfactorily completed the flight review required in §61.56(a) on *(date)*

_____.

SIGNED _____ DATE _____
CFI NO. _____ EXPIRATION DATE _____

I certify that *(First name, MI, Last name)* _____,
(pilot certificate) _____, *(certificate number)* _____,
has satisfactorily completed the flight review required in §61.56(a) on *(date)*

_____.

SIGNED _____ DATE _____
CFI NO. _____ EXPIRATION DATE _____